Henry Thompson

Clinical Lectures on Diseases of the Urinary Organs

Second Edition

Henry Thompson

Clinical Lectures on Diseases of the Urinary Organs
Second Edition

ISBN/EAN: 9783337018894

Printed in Europe, USA, Canada, Australia, Japan

Cover: Foto ©berggeist007 / pixelio.de

More available books at **www.hansebooks.com**

CLINICAL LECTURES

ON

DISEASES

OF THE

URINARY ORGANS.

CLINICAL LECTURES

ON

DISEASES

OF THE

URINARY ORGANS.

DELIVERED AT UNIVERSITY COLLEGE HOSPITAL.

BY

SIR HENRY THOMPSON,

SURGEON-EXTRAORDINARY TO H.M. THE KING OF THE BELGIANS; PROFESSOR OF
CLINICAL SURGERY, AND SURGEON TO UNIVERSITY COLLEGE HOSPITAL.

SECOND EDITION.

LONDON:
JOHN CHURCHILL & SONS, NEW BURLINGTON STREET.
MDCCCLXIX.

LONDON:

PRINTED BY WOODFALL AND KINDER,

MILFORD LANE, STRAND, W.C.

LIST OF LECTURES.

PREFACE.

I THINK it right to say that these Lectures were never committed to writing by me. They were delivered in a colloquial style, after the arrangement of the subject had been well considered, and were reported verbatim by one of our best shorthand writers. The copy furnished by him was corrected, some of those tautologies which seem to be necessary in teaching removed, and then sent to the *Lancet*. But each Lecture still required more space than was available in the columns of a weekly journal, and I further reduced it, perhaps one-fourth. I now present, in one small volume, at the suggestion of, I may truly say, numerous correspondents, known and unknown to me, the corrected copy in full, unchanged in form, and therefore unshorn of the familiarities which the conversational style peculiar—and, I believe, appropriate—to the class-room demands. And I do this, also, because I prefer that these Lectures, originally short, should not suffer any abbreviation, and because I desire to offer, not merely to the members of my own clinical class, but to students at large, some of the fruit of a long and careful study in that field of practical medicine, in its widest sense, to which they relate.

35, *Wimpole Street, London.*

PREFACE TO THE SECOND EDITION.

THE first edition having been exhausted in less than twelve months, the second is almost a reprint of the first, none but very slight alterations having been found necessary.

October, 1869.

DISEASES

OF

THE URINARY ORGANS.

LECTURE I.

INTRODUCTORY: DIAGNOSIS.

GENTLEMEN,—I propose to give a course of lectures on the Surgical Diseases of the Urinary Organs, and my object will be to afford you that information which will be most useful at the bedside. I shall not consider their anatomy or physiology, since that would make the course much too long. In the systematic lectures of the College it is impossible to communicate all those practical manœuvres, those little things which one arrives at by experience, either in the way of diagnosis or treatment, which are so valuable in practice hereafter ; neither is it possible that you should all acquire them at the bedside, since no hospital can furnish patients sufficient for the purpose; but you can learn a great deal by the conversational communications which are made here. It will be my aim to render available the result of experience which it has cost me years to acquire, and I shall do my best to furnish to you what of it is thus communicable in as many hours.

I have selected this course of clinical lectures on the

urinary organs for two reasons.   First, because my wards always afford *groups* of these cases; you can always find there abundant material for consideration at the weekly clinical lecture.   Secondly, because I do not know any set of diseases that are so successfully dealt with if you understand what you are about, or any in which you may make such dangerous mistakes if you are not well acquainted with them.   Neither do I know any diseases in which you can afford so much relief to suffering, none in which a skilled hand can do so much for the patient, and none in which you can gain more credit for yourselves.   It is therefore exceedingly important that you should be thoroughly acquainted with them.

I hope, in the course of about twelve lectures, to carry you through the greater portion of the list of subjects named in Part I.

I. Diseases of the Urinary Passages.

a. Diseases essentially inflammatory.

Urethritis,
Prostatitis, } acute and chronic.
Cystitis,

b. Diseases essentially obstructive.

Stricture of the urethra.

Hypertrophy of the prostate.

c. Calculous diseases.

Of the urethra.

Of the prostate.

Of the bladder.

Of the pelvis of the kidney.

d. Tumours—malignant and non-malignant.

Of the prostate.

Of the bladder.

II. Diseases affecting the Secreting Organs.

All organic changes in the kidney; also those altered conditions of the urine which depend on constitutional disease, such as Bright's disease and saccharine diabetes.

But, before doing so, I shall ask you to consider for a moment the title I have affixed to this course—viz. "The Surgical Diseases of the Urinary Organs." Now, you may inquire, "What are the surgical diseases of the urinary organs, and what are not?" To my mind it is very easy to tell you what are surgical diseases of those organs, but not so easy to tell you what are not. Look at the list before you, and see where the line should be drawn. Certainly the first division belongs wholly to that class—all diseases of the urinary *passages*, excluding the kidneys, which we will assume to be secreting organs. Undoubtedly all that part belongs to the surgeon. The physician, conventionally, claims the second; but since it is impossible to make a diagnosis of any one of those diseases without well understanding the whole, and as the physician does not make a physical examination by means of an instrument, I am compelled to regard all affections of the urinary organs as naturally coming within the province of surgery. This statement may not be universally received; but if we consider the matter, we shall see that it is absolutely necessary for the diagnosis of urinary diseases to be able to pass a sound or catheter. I do not say that the physician is incapable of doing this; but, conventionally, it is not practised by him. And you can no more treat diseases of the urinary organs without the ability to use these instruments than you can treat diseases of the chest without understanding the use of the stethoscope.

The first step in our course is naturally that relating to Diagnosis. I say almost nothing about the pathology and treatment of any one of these diseases to-day. The question before us now is diagnosis, and you will understand this to be a most important thing in all diseases—to know accurately what you are about to treat: there is then little difficulty as to the management. Many books can tell you the one; no book can tell you the other. Diagnosis can only be accomplished by the application of certain rules after some practice. It is the first thing to learn and to use; it is the last thing to be perfectly acquired. Indeed no man, let him live as long as he may, will ever be a perfect diagnostician. He may approach perfection; but if he is a diligent student, as he ought always to be, he will improve his powers of diagnosis as long as he lives. That is the reason why age or experience gives value to an opinion. It is long observation and extended experience that enable a man to arrive at diagnosis with greater certainty than the younger practitioner can possibly do.

Then we want to learn not merely diagnosis, but the *art of making a rapid diagnosis*. When called to the bedside, your action must often depend on the first three or four minutes of your interview. It may be easy to go home quietly, think over the case, pull down the authorities, and say, "I think the patient has so-and-so." That will not always do : it may do in some cases, and it had better do than that you should attempt to treat the case without having made up your mind as to the diagnosis. But that which will make you successful, that which distinguishes between the intelligent practitioner and him who is not so, is the ability to make a rapid as well as an accurate diagnosis of the case before him. Now, in calling your attention to

this subject, I am afraid I cannot guarantee that you shall
leave this theatre an hour hence first-rate diagnosticians of
these diseases ; but I can give you the method which, after
a good deal of thought and experience, I have found best to
answer the end proposed, and you can apply it afterwards
for yourselves.

First, you should pursue your diagnosis on a uniform
plan—that is, you should adopt a uniform mode of inter-
rogating each case of urinary disease. And what I say of
this disease is applicable to most others. Your object is to
collect *facts*, and your diagnosis consists of the *inference*
which you draw from those facts. You will endeavour to
arrive at the facts by the shortest possible route, and by the
most accurate method. You should ask questions, make
observations by the eye, by the hand, and by instruments,
and then examine the secretions. Take them in that order:
first, observation by questions; second, observation by the
eye, by the hand, and by the instrument, which after all is
but a long finger. You have no finger which is long enough
to go down these narrow passages, and you lengthen it by
means of your instrument. So with regard to the eye: your
endoscope, whatever it may be worth (as to which I shall
have something to say presently), is simply an increased
ability to see.

First of all with regard to questions. You may make out
most cases of urinary disease—say five out of six—by four
simple questions, including the minor extensions which arise
out of them. I always ask the patient these four questions,
and in the following order : —

The first question is, " Have you any, and, if any, what
frequency in passing water ? " Then, as a branch of that
question, springing out of it, I ask whether the frequency is

more by day or by night, or influenced by any particular circumstance. How the question applies I will tell you afterwards.

Then, secondly, I ask whether there is pain in passing urine, and whether *before*, or *after*, or *during* micturition; and whether at other times also, and if produced or aggravated by quick movements of the body. The locality of the pain is also to be ascertained.

Then I ask, as a third question, " Is the character of the urine altered in appearance?   Is it turbid or clear?" Possibly the patient will tell you that it is turbid; but you find, on questioning further, that it was passed perfectly clear, and only became thick after cooling or standing. Also, as arising out of this, you may often ask, "Does it vary much in quantity?" noting, of course, the specific gravity. The healthy standard, however, must be allowed very extensible limits; still quantity, I need not tell you, is a very important element in regard of renal disease.

The fourth and last question is—whether blood has passed in any way, whether it is florid or dark, whether passed at the end or at the beginning of making water, or whether independently of micturition altogether; and such like supplementary inquiries.

These are the four questions; and let me remark, that the answers you get will depend very much upon the way in which you put the questions. The patient is not always self-possessed, or he does not clearly understand the nature of the question you put.   It is necessary to be very precise and very distinct in your questions if you wish to get accurate answers.   In fact, there is no such difficult thing in all experience, whether in our profession or out of it, as to arrive at facts; and let me remind you again, that diagnosis consists in the acquisition of *facts*, and that it is impossible without them. Now you will say, how do I apply this to the list of diseases before you ?

First, as to the frequency of passing water. There is no serious affection of the urinary organs, except one or two which I will name, in which you have not more or less frequency of passing water. Thus the following is an exception: A man may have stricture to a considerable extent; the stream may be very narrow, and he may not for some years complain of frequency of passing water. Now observe, I have classified these diseases, so that we may deal with them more easily. First of all, there are the inflammatory diseases—inflammation of the urethra, of the prostate, and of the bladder. In all these you have frequency in passing water. Not necessarily, however, in urethritis, until it reaches the distant part of the canal near the bladder. I do not propose to enter upon the subject of urethritis here, as you have frequent opportunities of studying it in the out-patients' room. I am now only referring to this symptom of frequency of passing water as existing more or less in all these three diseases at some time or another. In hypertrophy of the prostate you have it, and it is remarkable that it is more at night than in the day. In chronic prostatitis it is usually present to a small extent; in cystitis it is, of course, a characteristic symptom. In calculous diseases it is prominently met with, and generally its degree is in proportion to the amount of movement permitted to the patient. Tumours, of course, malignant and non-malignant, are attended by the same symptom. In pyelitis, and in almost all organic changes of the kidney, in Bright's disease, and in diabetes, there is frequency of making water. Whenever the natural characters of the urine are altered before it reaches the bladder, the secretion produces irritation. This fact is worth dwelling upon for a moment. Diluted or watery urine is often regarded as unirritating ; on the con trary, it is not generally well retained by the bladder. The bladder is, as a rule, never so content as when it contains a urine of average, or more than average, specific gravity. Some persons, hysterical patients for example, will pass urine which is quite pale, almost like natural water, and the bladder is always more or less irritated by it. Of course, in diabetes, you have not only the character of the urine altered, but the quantity much increased. And I may remark that it is chiefly in renal affections that increase in quantity takes place ; while, on the other hand, suppression of urine is always a malady of the kidneys.

The next question has reference to pain ; and when you get answers as to the nature and seat of pain, you will begin to see your way towards a diagnosis. In prostatitis there is usually pain at the end of passing water—less severe, but resembling somewhat that of stone ; as the bladder con-tracts, when empty, on the tender prostate. In cystitis the pain is usually before micturition, because the inflamed bladder is sensitive on being distended, and is anxious to get rid of its contents. The pain is just above the pubes. When cystitis is acute, pain may be felt in the perineum also ; but in chronic or subacute cystitis it is supra-pubic, and not at the end but at the beginning of making water, unless the prostate is affected, and then the tender prostate gives a little pain at the end, as I have just said.

In stricture of the urethra there is often pain about the seat of the obstruction, an idea of which you may obtain by a simple experiment. If, when passing urine with a full stream, you suddenly narrow the passage with your finger, so as to diminish the stream one-half or more, you will experience an acute pain.

There may be pain with hypertrophy of the prostate, inasmuch as this is frequently associated with chronic cystitis, when the pain is before making water, and not afterwards—differing in that respect from stone. The bladder wants to get rid of its contents, and can do so but slowly, on account of the enlarged prostate, which stands as a barrier in the way. During its first contractions, which expel but little urine, there is pain above the pubes and deep in the perineum; but when a third or a half of the contents has issued, the patient is relieved.

I shall not dwell upon calculous disease of the urethra. The calculus is only a temporary lodger there, and as it can often be felt externally by the hand, there is rarely any difficulty about the diagnosis. Calculous disease of the prostate is also rare. I shall not complicate what I wish to be a simple matter by dwelling upon it, but call your attention to the commoner condition of calculus in the bladder. Here the pain is quite distinct in its character: it is felt at the end of passing water, because, the bladder being emptied, the rough surface of the stone is left in contact with the mucous membrane, doubtless that covering the neck of the bladder, which is unquestionably a sensitive spot. As soon as sufficient urine has trickled down into the bladder to separate the coats from the stone, relief is obtained. Then the pain is felt at the end of the penis, within an inch of it, about the base of the glans. Furthermore, the pain is increased by movement: in other complaints it is not necessarily so. Put a patient in a rough-going vehicle, or make him jump from a step, or perform any rapid movement, and instantly he feels severe pain, probably at the neck of the bladder, but also at the end of the penis. In prostatitis, inasmuch as the neck of the bladder is involved, there is usually some pain at the end of the penis, which is a reason why chronic inflammation of the prostate is sometimes mistaken for stone.

With regard to calculus of the kidney, I have little to say about it here. Of course you have pain referred to the locality, right or left, not often to both kidneys; there is tenderness also, and much increase of pain on movement. It is usually on one side only, and perhaps more frequently on the left than on the right side.

One cannot, perhaps, say much about any characteristic pain in connection with tumours. They may be situated in any part of the bladder; may obstruct the urine more or less; and accordingly as they produce cystitis, and obstruct the flow of urine, pain will be experienced.

The next question is as to the character of the urine itself. Now, suppose your patient has told you that he has frequency in passing water, pain at the end of the penis and at the neck of the bladder, and that the pain and frequency are aggravated by movement. You may begin to imagine, "Perhaps the man has stone in the bladder, and I shall have to sound him." Two questions only have already put this probability in your way, and you interrogate as to the character of the urine. See how this carries you a step further. We recommence our list as to this inquiry. A preliminary remark, however, about examining urine. I do not propose to teach you here a systematic mode of doing this. It is not in my department, and would only be repeating that which it will be your duty to learn elsewhere, and I hope you will do so thoroughly. But there is this hint which I may give with respect to it. Whenever you

want a specimen from your patient to examine, do not tell
him to send you a bottle of it passed in the usual way, or
you will get a mixture of often doubtful value. What you
require is the secretion of the kidneys, plus only anything
there may be in the bladder; you do not want it complicated with anything which may come from the urethra. Let
the man pass two or three tablespoonfuls through the urethra
first, so as to sweep out whatever may be there, which may
be put into a separate bottle, and then you will get a specimen—at any rate one of which you will know the source.
You will have the renal secretion plus anything in the bladder. Suppose the man has gleet or chronic prostatitis : there
will then be a quantity of muco-purulent matter in the
urethra. If all this be carried into one vessel with the
urine, how will you determine the different products, and
decide, by the eye or by the microscope, what has come from
the urethra, what from the prostate, and what from the kidneys ? You cannot do it; but if you get rid of the source
of error by flushing the urethra, so to speak, and emptying
the contents into a separate glass—say a wineglass—then
taking the bulk or remainder in a tumbler, you will generally have a sample of urine that you can do something with.
If I felt disposed to indulge you with gossip, I could tell
you stories of the gravest blunders committed by not attend-
ing to that simple point. I can at all events tell you that I
have more than once known a learned practitioner treat a
patient for pyelitis who had nothing but a profuse discharge
from the urethra ; how the urine had been sent twice a week
in a bottle scrupulously made clean for the purpose ; and
because a quantity of pus was found in it, the patient, who
had some symptoms corroborating that view, was treated
during some months for pyelitis; how a surgeon at length

found out that the whole of the matter came from the
urethra, so that when the urethra was flushed into the first
glass all the matter was there, and that the remaining urine
was clear and healthy; and, finally, that the "pyelitis"
soon disappeared under local treatment of the urethra. I
do not know whether any one else may tell you of that simple
mode of determining this matter; and I will assume that in
the future you will none of you make such a mistake as that
I have mentioned. I only know too well how necessary it
is to call attention to it, and how seldom it is done.*

Referring first to prostatitis, it is always associated more
or less with shreds in the urine, which come from the prostatic part; and if you separate the urine as I have told you,
you will find that the whole of the thick matter will be in
the first glass, while that remaining behind will be clear.
How would it be with regard to calculus? You might have
muco-pus in the first glass, but you would have more in the
second from the bladder. Not very often have you calculus
in the bladder without having some muco-pus from the
bladder itself. Occasionally, but very rarely, do I find a
man with stone in the bladder having perfectly clear urine.
Not commonly do I sound a man for stone who has clear
urine, unless he has marked symptoms, because the presence
of stone in the bladder almost always gives rise to a certain
amount of cystitis, and there is deposit in consequence. If
the patient passes shreds of thick matter in the first glass,
and the urine left behind is clear, and has symptoms like
those of stone, rely upon it it is a case of chronic prostatitis.

The character of the urine in one of the forms of chronic
cystitis is well known. There is at the bottom of the vessel

---

* See further remarks on this subject at the close of the last lecture.

a thick mucilaginous deposit, which does not issue in a
stream, but falls out in a mass. In acute cystitis the urine
is cloudy, and there is a considerable deposit of pus. In
stricture of the urethra, unless chronic cystitis has been set
up, there is no deposit from the urine. Here the character
of the stream is important. If, when the patient is passing
urine, you see a very thin, small, spluttering stream, or
urine issuing only in drops, you will know that there is an
obstruction, most likely stricture; because, although in
hypertrophy of the prostate you may have the stream much
diminished, it will be a stream which falls downwards from
the organ. In stricture force may be brought to bear on
the stream, so that, however small it may be, it is often
fairly propelled; but in hypertrophy, in which the expelling
apparatus is involved, the muscles cannot act, and therefore,
however large or small the stream, it generally falls perpendicularly. In calculous disease of the bladder there is
nothing to note about the nature of the stream, except that
it stops suddenly sometimes—by no means a constant symptom. With regard to the débris of tumours found in the
urine, the microscope sometimes, not often, throws light
upon their nature. No doubt you may see cancer-cells in
the urine, but it is difficult to identify them. I have seen
such cells declared to exist by good observers in cases in
which cancer was not present. Young pavement epithelium
is easily mistaken for them. Going upwards from the bladder,
we may note pyelitis, more or less chronic—a disease in
which the condition of the urine is only one symptom
among many others which must be observed before arriving
at a conclusion.

The next question is, "Do you pass blood?" and this
will bring you very near indeed to an opinion on most cases

—not quite, because in any case you may first have to sound. In prostatitis there is often a little blood at the end of micturition, as in stone; in cystitis there is not necessarily blood, unless it is acute and far advanced; in stricture of the urethra there is not necessarily blood; and in hypertrophy of the prostate not necessarily. You may have it or not, often only as the result of instruments. It tells most in the question of stone. Just as in phthisis a large proportion of patients have hæmoptysis at some time or another; so in about the same proportion of cases—say four out of five—there is some blood with vesical calculus.

I want you to pay particular attention to these questions, because I shall assume an acquaintance with them to underlie much of what I have to say hereafter. What I wish to add with regard to observation by the eye, by the hand, and by instruments will come under each particular subject hereafter, and I will only briefly allude to it to-day. By the eye you observe mainly whether the bladder is distended or not, and you are assisted in ascertaining this by palpation and percussion. You examine the perineum and scrotum also, with a view to extravasation of urine, perineal abscess and fistula, &c. And now we come to the question of instruments. Suppose such a case as that to which I have already referred, in which there are frequency of passing urine, pain at the end of micturition, pain on any considerable movement, thickening of the urine, blood passing occasionally, but more on movement—you regard it as highly probable that the man has stone. You cannot arrive at a certainty without instruments. You may have almost all these conditions in certain changes of the kidney and in renal calculus, and you cannot distinguish them unless you skilfully explore the bladder with a sound. When I claim great value for this instru-

ment, quite understand that I am by no means desirous that
in the case of every patient who comes to you and com-
plains of some frequency in making water, or pain in the
act, you should say, "Lie down, and let me pass an instrument." Perhaps the surgeon may be apt to abuse a little
his power of passing instruments : it should never be done
unless it is absolutely necessary. I hold that an instrument,
per se, is an evil—a very small one or a considerable one,
according to the manner in which it is employed—and that
it is never to be used unless there is good reason to believe
that a greater evil is present which it may mitigate or cure.
But when your patient has the symptoms named, you will
be doing him an injury unless you resort to it. In cases
of stricture the instrument is also necessary; and so in
ascertaining the condition of the bladder, whether it is full
or empty. A man may make water very frequently, strain
hard, and be very certain that he has emptied the bladder,
and be quite deceived. How can you tell? You find a
prominence above the pubes which you have no doubt is a
distended bladder; but it is just possible that it may be a
solid tumour. You cannot know whether the bladder is
emptied unless you pass an instrument. Many a man has
had an instrument passed into the bladder, and a quart of
urine has been found to be left behind, when his own sensations led him to believe that he had expelled every drop.
We shall see more of this when we come to the subjects of
retention and hypertrophy of the prostate.

As we are now on the question of diagnosis, I take the
opportunity of showing you that the eye may be assisted to
a certain extent by what is called the endoscope, which is
simply an instrument that we have long been in the habit
of passing into all the cavities of the body—the ear, the

vagina, the rectum—for the purpose of bringing reflected
light to bear upon the interior of those cavities. For some
years past this instrument has been employed for the
urethra. It is seventeen years ago since I first saw the
endoscope so applied. This was in the hands of Mr. Avery,
of the Charing Cross Hospital. As I was turning my attention somewhat to this subject, he asked me to see some of
his patients, and a new instrument he was then making.
He showed me a long tube, precisely similar to this which
I hold in my hand, with certain arrangements enabling one
to see deep portions of the urethra. He showed me cases
of stricture, but I do not think he looked into the bladder.
He paid a great deal of attention to the subject, and the
instrument was brought by him to a certain state of perfection ; unhappily, however, his death occurred shortly afterwards, and the thing was lost sight of here. Various
attempts have been made with the same object, long before
and since, but I do not know that there is anything on this
table very superior to what Mr. Avery showed. Within the
last few years M. Desormeaux, of Paris, has paid great
attention to the endoscope, and has perfected one of his
own, consisting of a similar tube, but with different appli-
ances. The various modes in which light is applied constitute the differences between the various kinds of endoscope.
In all of them there is a tube of this description to pass into
the cavity. Six years ago I had an endoscope of M. Desormeaux's, and exhibited it in the hospital—the instrument
which you see here. Dr. Cruise, of Dublin, has brought it
to greater perfection, and has produced a better instrument
than we heretofore possessed. This also is here, and you
have often seen it in the wards, applied by me both to the
urethra and to the rectum. I may tell you at once that if a

man has a good and a tolerably practised hand, with a fair share of intelligence, I do not think he will gain a great deal by the endoscope; and if he has not, I think it will be of no use at all. There are some few cases in which he may find it of value; but do not expect that the endoscope is going to work any marvels in the diagnosis of surgical diseases of the urinary organs. In nineteen cases out of twenty you ought to be able to arrive at the necessary information without it. And it is not the easiest thing in the world to apply. As already remarked, a man should not be put unnecessarily to the pain and inconvenience of a sound or a catheter; but examination by the endoscope is a somewhat more irritating and tedious process. In certain exceptional cases, in which you are unable to arrive at a conclusion without it, you may employ it to some advantage. Now, here is a patient on whom I have never used it, and whose case will offer a certain test of its power. The man before you had an exceedingly bad stricture of the urethra, which I cut internally last Tuesday week. He is now perfectly well. He could not pass a drop of urine before the operation, but now he is able to pass it naturally; and you will agree with me that a great deal must have been done last Wednesday week to make that change. I cut through the strictures deeply, and now we shall see whether we can find the cicatrices. I shall use Desormeaux's endoscope, illuminated by Dr. Cruise's lamp.—You see we have now made a careful and prolonged examination. The urethra is of a more dusky red about the part which has been affected, but that is all which can be observed. Changes in the colour and texture of the mucous membrane of the urethra and bladder are those which are most easily seen, and which are of the most importance to note. The orifice of a stricture may be some-

C

times seen, but the result is without practical utility. A
stone in the bladder may be easily seen, or rather the small
portion of it upon which the end of the sound impinges;
but I have never gained anything by the sight. A calculus
smaller than a pea may be easily found by delicate sounding,
and an audible note elicited from it, more easily than you
can see it through the endoscopic sound. I may mention
that no one has yet been able by its means to identify the
veru montanum, and if you cannot see the veru montanum,
I think it is quite possible that minute pathological changes
will often escape you.

Mr. Baker has been good enough to send up another and
very simple endoscope, designed by Mr. Warwick, for exhibition. It may be used with ordinary gaslight or with sun-
light. It certainly seems, on comparison, to effect nearly
as much as the larger and more elaborate instrument.

LECTURE II.

STRICTURE OF THE URETHRA.

LAST week, gentlemen, the subject of the lecture was rather general than special. In commencing this course, I shall take to-day the subject of stricture; and I do so because, if not really one of the most common of these disorders, it is often supposed to be so. Among the many complaints of this class respecting which you may be consulted, perhaps none will be more talked of than urethral stricture. It does not follow, however, that stricture is really so common ; in fact, it is much less so than many suppose. The word happens to have been popularized, and therefore, when a person experiences a little trouble in passing water, he is very apt to say that he has stricture. Certainly, in three out of four cases in which persons do so, I find there is nothing of the kind, but often merely some temporary cause of irritation. Then it must be confessed that there is some confusion, even amongst the profession, as to the mode in which the word "stricture" should be employed. It is said—and I have said it myself, because I originally adopted the usual classification—that there are three kinds of stricture—organic, inflammatory, and spasmodic strictures. Now, it would save some confusion if we employed this word for only one kind—namely, organic stricture. And what is organic stricture ? It is a deposit of lymph round the canal of the urethra at some point, which, not allowing the canal to open to the stream, narrows the current *pro tanto*. There

has usually been some chronic inflammation, most commonly
in the bulbous part of the canal, and a deposit of lymph has
taken place in the submucous and in the vascular tissues
surrounding the urethra; and after this lymph has existed
some time, it contracts and forms fibrous bands more or less
encircling the canal. We talk of the *contraction* of the
canal; but this is due to a popular and not very correct
notion of the matter, although it answers well enough for all
practical purposes. You will do well to remember, in connection with the pathology and treatment of urethral diseases,
that the urethra is not an open tube, like a gas-pipe, for
example, into which you can pour fluid. On the contrary,
it is always absolutely closed by muscular fibres, and it is
only when the tube is habitually prevented from fully dilating
to the stream of urine that it is strictured.

And this organic stricture is a permanent condition. Once
acquired, it cannot be dissipated by any known means. It
cannot be removed by absorption. You may dilate it, you
may cut through it, but there it will always be. When a
man once has true organic stricture, he has it for ever. If
any exceptions exist, the rarity is so extreme as practically
not to invalidate the axiom laid down. Whatever treatment
you employ, there is always a tendency more or less to contract afterwards.

Now touching "inflammatory stricture" and " spasmodic
stricture." Inflammatory stricture is merely a temporary
local inflammation of some part of the canal, which is then
narrowed for the time. The patient is unable, as long as
that inflammation lasts, to pass water, or at best with difficulty. The inflammation usually affects the prostatic part
of the urethra, which is not, as you know, the seat of
organic stricture. If you call this condition stricture,

you may as well say that the throat is strictured when it is inflamed and the tonsils are swollen. We only speak of stricture of the œsophagus or gullet in reference to a condition which is organic, when by some deposit the passage is permanently narrowed, and we never speak of stricture there under any other circumstances. So with regard to spasm. The urethra may be narrowed to a certain extent by spasm —that is to say, the water may be prevented from passing outwards from the bladder, because there is some irregular action of the muscles around it. But it is only temporary; it does not necessarily imply any organic change; although sometimes its occurrence depends on the existence of organic change, yet this spasm is not stricture of the canal. I will tell you what spasmodic stricture is. It is exceedingly useful as an excuse for the failure of instruments. It is a "refuge for incompetence." When you cannot pass a catheter, when you find it exceedingly difficult to get anything in, and in fact wish to desist, it is a convenient thing, and has always been so recognized, for the doctor to say, "There is spasm." Indeed, I believe he often "lays the flattering unction to his soul" that it does exist, although, in my opinion, it does not, or at least very rarely. "There is spasm," says he, "now in the muscles, and it will be prudent at present to desist from further attempts to pass an instrument." And no doubt when this is said it is so. Now, I do not think that you ought ever to fail in passing an instrument because there is spasm. Spasm may prevent the urine from going outwards; I do not know that it ever prevents the instrument from going in. In most cases it is failure of the hand, not spasm of the urethra. Still I cannot deny that it is a useful excuse—that it has a sort of foundation in fact, and may thus be often a better explana-

tion for the patient than anything else, when the instrument
does not pass.   But when we speak here of stricture in
future, we shall refer only to organic stricture, in the sense
already described.   All the mechanical treatment which I
shall have to speak of will have reference only to that kind
of stricture.   In "inflammatory stricture," of course, you
have no occasion for instruments, unless retention of urine
is present.

Now, what are the symptoms of stricture?   First, of
course, there is the smallness of the stream depending upon
the narrowed state of the canal.   Whatever the narrowing
of the canal is, in that proportion there must be a narrowing of the stream.   Next, there is often some straining to
pass water, corresponding to the obstruction of the passage;
and the stream itself is flattened, twisted, or divided.   This is
accompanied by pain at the seat of stricture, and sometimes
also over the pubes, if there is any cystitis.   Associated
with this, it is common to have a little discharge from the
urethra; indeed a gleet is often the only thing which the
patient notices at first, and the surgeon, finding that this is
not readily cured, uses an instrument and discovers stricture.
Frequency of making water, as I told you last time, is not
always present in stricture, although it always is so when
the case is severe and of long standing.

Supposing a patient to apply to you with all these symptoms, you endeavour to see him pass water.   He will probably lay some stress on the fact that it is twisted or divided.
Do not attribute much weight to this circumstance by itself,
for a twisted stream often occurs when there is no stricture.
It may be due to an alteration of the external meatus; for
as the stream issues from the passage it is modified by the
shape or rigidity of the external meatus, and after repeated

inflammation there, the lips of the meatus may be slightly thickened, and permit only the exit of urine in a flattened and consequently twisted stream; and this is by no means an uncommon occurrence.

We now come to the important point of diagnosis. One rule is to be invariably followed : whenever you examine for stricture, take a fair-sized instrument, say No. 8 or 9. The patient may remonstrate, and will probably say, "That won't pass; it is useless to employ so large an instrument." Tell him that you do not propose to pass it, but only to discover where the obstruction is. For if you use a small instrument at first, it may pass through the stricture, if one exists, without detecting it; but if the large instrument goes on easily into the bladder, you have the satisfaction of telling your patient that he has no stricture, and you look further for the cause of the difficulty. Whatever, then, a man may tell you, and however small the stream may be, take an instrument not less than No. 8 or 9, pass it gently down the canal, and if there is a stricture the instrument is arrested, and you will find the exact position of the obstruction. Notice the point where it exists—four inches, or five, as the case may be, because it will be useful afterwards in dilating to know its exact locality.

Now, in passing the instrument, you may meet with cir- cumstances which may mislead you. I have spoken to you of error on the part of the patient, and I am bound to say that the surgeon who is not much practised in these matters may also be deceived. What are the sources of fallacy to which he is exposed ? How is it that he sometimes fancies there is stricture when there is not ? There are some kinds of practice in which you may be hereafter placed, that do not afford the opportunity of often seeing this disease, and

in such it is no great discredit to a man to think that he
has found stricture when none exists. Not, of course, if
he is a professed surgeon ; then it will be a great discredit
to him. But if he has very little to do with this kind of
thing, he may encounter some difficulty in the urethra, and
he may suppose, but erroneously, that it is due to stricture.
Now, I want to guard you all against this ; for, though you
may not all be operating surgeons, I want you not to leave
any course of lectures which I may give, without knowing
precisely what are the sources of fallacy, so that I may not
hear of any of you hereafter making such a mistake as that
to which I have referred. Let me remind you again of
what was said about the urethra being a closed canal. I
am not going to draw you a diagram of the urethra such as
you see in anatomical books, in which it is represented in
section as an open passage ; for it is never in that condition
except in the act of micturition. First, close to the meatus
is a source of fallacy—I mean the lacuna magna. Then,
here, five or six inches further on [a diagram referred to],
the bulb joins the membranous portion, and the canal, from
being wide or dilatable, becomes less so. Lastly, there is a
source of difficulty at the neck of the bladder. Those are
the three points at which persons may be mistaken in passing
an instrument, and form erroneous notions in consequence
respecting the presence of stricture.

Now, whenever you pass an instrument, do not let your
thoughts revert to those anatomical diagrams representing
the urethra as an open tube ; bear in mind that it is simply
a closed passage ; so that nothing is easier in passing an
instrument, if you are out of the line, than to find some
obstruction in the folds or lacunæ of the mucous membrane,
since it has more distensibility at some points than at others.

Of course, if it were simply an open tube, there would be less difficulty in going on ; but, as it is not so, the point is liable to hitch against the soft parts on one side or the other. First, as I have said, it is quite possible to get stopped at the very outset, which is embarrassing to a beginner, by engaging the point of the instrument in the lacuna magna. Whenever, then, you begin to pass an instrument, let your first thought be to keep its point on the floor, so as to avoid that obstacle. You wish, of course, to pass it well for the patient. Perhaps he has had instruments passed before, and you desire to succeed at least as well as the preceding operator. Now, there is nothing which a patient appreciates so much as the easy passing of an instrument. It is a disagreeable operation, and if you pass it more easily than other persons, you will probably retain your patient as long as he requires assistance of that kind. If you stick and get into the lacuna magna at the outset, he infers you to be a bungler, and perhaps will not come to you again.

Now, you see represented in this diagram the bulb of the

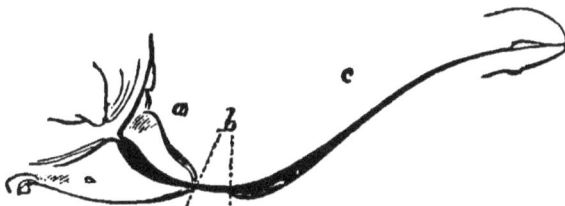

Fig. 1.—Diagram of urethra in natural condition, *a*, *b*, and *c* representing the prostatic, membranous, and spongy portions respectively.*

urethra. The canal is more distensible at this point, and when you come to the deep perineal fascia the canal is much

* The urethra should have been shown here as it really is, a closed canal ; the line in the bulbous and prostatic portions having been made by me merely a little thicker to mark (*diagramatically*) position and the character of dilatability. This line has been somewhat exaggerated by the artist.

less distensible. Practically, therefore, it is much wider there than it is at the orifice, and when you get the instrument at that point it is apt to hitch. This is the place where almost all the false passages are made: the instrument is driven out of the canal down below the urethra, it being mainly at the floor that the canal is so distensible. The section of the corpus spongiosum is wider below than above; the texture is soft and spongy. The urethra corresponds in distensibility to the soft structure outside, and although the instrument gets smoothly down to this point, it may not go on into the membraneous portion. Now, take care, at first, to get the point of the instrument so turned up as to avoid this lower part. Nothing is so good as a well-curved instrument to avoid that obstacle. I am in the habit of doing this in the case of out-patients. I like to take a student who has never passed an instrument before, and I say to him, " Pass this bougie (a straight or slightly curved one) into the canal." He passes it, and invariably, when he arrives at the membranous portion, stops. I then take the

Fig. 2.—Bougie, with point turned up.

same instrument, give its point this form (Fig. 2), and then he passes it immediately into the bladder. As the instrument goes in, it keeps close to the roof, instead of engaging itself in the distensible part of the bulb.

The last obstacle is at the neck of the bladder, and so

common is it, that you often hear of " stricture of the neck of the bladder"—a thing which never exists. There never was a stricture even in the prostatic portion. " Stricture at the neck of the bladder " was a household word some years ago, and even now you sometimes hear of it; but there is no such thing. It is simply because there is sometimes difficulty in passing the neck of the bladder that it came to be regarded as a locality of stricture. In this case also a well-curved instrument is the best thing to pass in. Let me recapitulate shortly the three sources of difficulty : First, the lacuna magna, which is avoided by keeping on the floor of the canal ; then the narrow membranous portion at the bulb, which is avoided by keeping the point of the instrument well up ; and the same with regard to the neck of the bladder.

I was going to read to you the notes of some cases of stricture now in the ward upstairs, but I will not do so, because you have seen the cases, and know pretty well what they are, and I shall only refer to some of them as we go on. One has been treated by continuous dilatation, and another by internal cutting. Then we have several patients treated by common dilatation, and it is of common dilatation that I have to speak to you to-day. Suppose that you have passed a full-sized instrument, and found that it stops unmistakably, say, at about five inches, the most usual place ; you will then, of course, have to pass smaller instruments, to see what will go through it. You have diagnosed the fact of the stricture, and the situation ; but you do not know the calibre. You ask the patient to make water. If he cannot do so, take a small gum-elastic instrument, say No. 1 or 2. You pass No. 1, and that perhaps goes through easily; then No. 2, which goes not quite so easily ; then No. 3, which goes in

tightly. No. 3, then, is the calibre of the stricture. Now
you have all the main facts about the stricture that you
want to know. The man's general condition, the frequency
with which he makes water, the amount of chronic cystitis
—all these are questions to be considered; but it is only
with the mechanical part that I have to deal to-day. Lastly,
we will assume that he has only one stricture; for a patient
may have two, but this is exceptional. Now, what is the
treatment? First and foremost, Dilatation—dilatation
always—dilatation without exception, whenever it will succeed. It is always to be tried first, because it is the simplest
and easiest mode. If you find a man with a stricture ever
so narrow or tight, by no means think of operating till you
have tried whether dilatation will succeed. What is dilatation? A mechanical process of stretching this organized
lymph, which forms bands round the canal at the strictured
point. It is often said to produce absorption of this tissue,
which I shall not deny, but only say that there is not the
smallest particle of proof to support the notion. Now, suppose you have been able to pass No. 3 bougie or catheter
rather tightly through the stricture and into the bladder, to
be perfectly sure that all has gone right, then you say to the
patient, " That is enough for to-day ; come again in two or
three days' time for a larger instrument." Now, I advise
you, on such second or third day, not to commence with the
largest instrument previously passed. Having passed Nos.
1, 2, and 3 on the first occasion, you should now take Nos.
2, 3, and 4 ; and on the third occasion, 3, 4, and 5 ; and so
on; always beginning below the point you had attained on
the previous occasion, making the smaller instrument a sort
of *avant courier* for the larger one. Further, never let the
instrument remain in the urethra; withdraw it at once;

leaving it there simply increases irritation, and does not augment in the slightest degree the dilating power. Thus you find that the longer you leave it the tighter it is held, and the more difficult and painful it is to withdraw. Not until it has remained in the stricture an hour or two does the stricture begin to relax, as we see shall by-and-by, in considering " Continuous Dilatation."

Now comes the question of the kind of instrument to be employed. The great principle which underlies all mechanical treatment of the urinary organs, whether for stricture or for hypertrophied prostate, for retention of the urine or for stone—the one great principle which must decide for us the question of the kind of instrument to be employed is this : *all instruments are to be considered evils, more or less, never to be resorted to unless a greater evil be present.* The passage of an instrument of any kind into the healthy urethra must *per se* be a source of irritation. Try it yourself; and I advise you to do it, if you wish to pass an instrument well ; for I hold that no man should pass an instrument for another until he has passed one for himself. Of course the amount of irritation will depend in great part on the manner in which it is passed, and on the kind of instrument employed. Let us consider, to use a commercial simile, that in the case of your patient there is a " debit and a credit side " in all treatment. You intend, beyond all doubt, to effect some real good—that is, to the " credit side " of the account; but you cannot do it without making some irritation in order to gain your end—that is an entry on the " debit side." Be careful, then, that you constantly bear in mind the latter fact, and make it your business to diminish that " debit " as much as possible. Do not pass an instrument unless there is some good reason, unless

there is some evil, for the sake of curing which it is worth
while to incur a little irritation. Acting upon this principle,
you will choose such an instrument as you know by experience or otherwise to produce the least possible irritation.
And this leads me to the question of the difference between
solid and flexible instruments. Here I feel that I am tread-
ing on delicate ground; and I will tell you why. First of
all, no one has been a greater advocate than I was some
years ago of solid instruments as against soft ones, influenced by the traditions of this place, which are entirely in
favour of the former. I can give you the reason for that.
The great master-spirit of this place, who has been dead
some twenty years or more, the man who gave the tone to
the place, and educated almost all the elder men here—I
mean Liston—declared his preference for the solid instrument
in very strong terms. It is (1867) just twenty-one years
ago since I sat in this room and heard him deliver a lecture
on that very subject. His powerful advocacy of the silver
instrument, and the contempt he had for others, were matters
of notoriety. Starting, then, with such views, and regarding
him—as every one does, to a certain extent, the man who
teaches him well and fairly what he learns—for a certain
time as an oracle, I was strongly in favour of the solid
instrument as against the flexible. But what is much more
valuable than any oracle, whoever he may be, is a large
personal experience; and this has taught me that, beyond
all question, the flexible instrument is the best—if only you
know how to use it—for the treatment of stricture, and for
all maladies of the canal, whenever it is available. I am so
certain of this, that I have no hesitation in saying that a
great part of the success of any man who has much to do
with this subject will depend upon his use of flexible instru-

ments as against solid. No patient will ever allow a surgeon to pass for him a solid instrument if you have passed for him a flexible one as easily as you are bound to do. It gives him so much less pain, and produces so much less irritation. To continue my commercial simile,—it puts so much less on the "debit side" of your patient's case; you get so much more of advantage, and so much less of disadvantage. I confess, then, to a considerable change of opinion even since I published my first work on the subject; and I do this without the slightest shame or the slightest repugnance. I hold that the end of life in this world is not achieved without change in our opinions. You may rely upon it, with regard to any subject whatever, whether politics, or religion, or our own proper profession, if we hold the same opinions at forty years of age as we did at twenty— and, perhaps, looking forward, I may say if we hold the same opinions at sixty as we do at forty—we live to very little purpose. It is an error to look for a life-long "consistency" in matters of opinion from men who think for themselves, in whatever department their teaching may be. You must expect them to progress, or they will be bad teachers —just as I hope you are all progressing now. I have said this because I know that so much might be quoted from what I held fifteen years ago in contradiction to what I am now saying. If I did not state this, you might ask me why, having said so much in favour of the silver instrument, do I now say so much in favour of the other. You have my reason; it is simply that I have learned better.

There are two kinds of flexible instruments, the English and the French. Inasmuch as the French instrument possesses more flexibility than the English, I often prefer it. Perhaps it is right that I should add a word or two to what

I have said. I believe the flexible instruments are much better now than they were in Liston's time ; and I think that, if we had had the good fortune to retain him so long (without taking credit for making a shrewd guess), he would have changed his opinion too. This is the kind of flexible instrument much used in his time. It is called a bougie ; and properly, since it is simply a kind of wax-candle, and is, in my opinion, a very imperfect instrument. You can bend it into any form by warming, but it is a very inferior instrument to what is generally used now. The gum-elastic or English flexible instrument is very valuable on account of one quality which it possesses, and which does not belong to the French instrument—*i. e.* it will preserve any curve, when cooled, which you choose to apply to it under the influence of heat. If I want a small curve, I take the instrument, put it in warm water, give it the curve required, then put it in cold water, and the curve is fixed or set. The French instrument is exceedingly flexible ; you may wind it round your finger without difficulty. And it has another valuable character—namely, its peculiar tapering point. Now, a tapering point *per se* is often undesirable in the urethra, because it is very liable to get into some lacuna. It would be an advantage if you could ensure that it did not do this ;

Fig. 3.—The bulbous-ended bougie and catheter.

but this liability is provided against now very ingeniously by means of a little bulb at the end (*see* Fig. 3). The long tapering extremity guarded by the bulb insinuates itself

through the healthy urethra, or through one not greatly contracted, in a most certain and easy manner. Such an instrument as that may be passed by the patient himself without difficulty. Indeed, it may almost be said to be " surgery
made too easy." The merest tyro can pass it in nine cases out
of ten, although he might not manage the tenth. It is one of
the most extraordinary instances of English conservatism that
these instruments are found in so few hands. They are, however, at last being made here. For years it has been necessary
to send to Paris for them ; but a demand is arising now, and
they are at length manufactured in this country. If you will
try the experiment on yourselves, you will find that this instrument traverses the urethra almost without your knowing
it; and it requires no knowledge of the canal in order to use
it. Now it may appear to you very heterodox, but I advise
you, in passing an instrument, to forget all about your anatomy. You are taught it over the way, and it is most important
that you should know it ; but, in passing an instrument,
forget all about the different regions. Think nothing about
the deep fascia, the membranous portion, or the compressor
urethræ. A solid instrument is especially dangerous in the
hands of an anatomist; he will push it the way he thinks
right, as if all urethras were exactly of the same form, and
did not vary as much as noses do, or other features. This
used to be the pretext for preferring the solid instrument:
it was said, "You want to know exactly your anatomy,
and pass the instrument accordingly." I pity the patient
who has a solid instrument thrust into his body by a knowing
man at anatomy. You want an instrument that you can
use most delicately, holding it lightly between the finger and
thumb, withdrawing it or changing its direction as soon as
you are able to perceive an obstruction. Your hand is

D

to be educated for the power of delicately perceiving the characters of the passage by means of the instrument within it; and rarely, if ever, are you to push a solid instrument in any given direction preconceived to be the right one. If you wish to combine the maximum amount of dilatation with the minimum amount of irritation, this flexible instrument is unquestionably the one for you to use.

There is one other point to notice with reference to French and English instruments,—I mean the gauges. Our numbers are from 1 to 12. Here is No. 12; and you generally consider, when you have reached that size, that you have completed the dilatation. In England, we cannot be said to have a uniform scale; all our measurements are very arbitrary. One maker has one scale, and another another; and the Scotch scale differs by one and a half from the English; so that the patient who takes No. 12 Scotch, takes only $10\frac{1}{2}$ English. Our more exact neighbours over the Channel use the millimetre, and the number represents the precise size, so that when I have named the size, I have named the

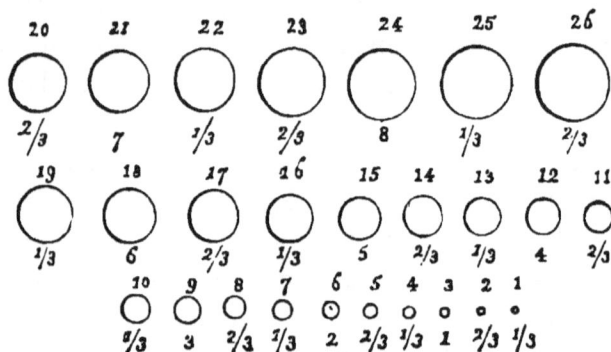

Fig. 4.—French gauge.

exact calibre or magnitude of the urethra. Instead of 1 to 12, the French have 1 to 30 (see Fig. 4). They begin lower, and go higher than we do, and the steps are more gradual. In this way irritation may be lessened. You may pass, for instance, a No. 4 English very easily, and a No. 5 with great difficulty, or not at all, and an intermediate one might be the proper size. Their numbers 3 to 21 are about equal to our 1 to 12; showing you at once how much more gradual the range is. No. 1 is one millimetre in circumference; No. 2 two millimetres, and so on; so that the increase in size is uniform as well as gradual. If I have a patient who will take No. 21, I know that his urethra admits an instrument twenty-one millimetres in circumference, and of course seven in diameter. I advise you in this, as in other matters, to be cosmopolitan in your views, and to adopt improvements from all quarters.

I have told you that simple dilatation consists in passing every two or three days a larger instrument, until you reach the highest. In many cases all goes on smoothly from the beginning to the end. Then you teach the patient to pass the instrument for himself, and he does so afterwards once a month, or once in six weeks, to maintain a sufficient calibre.

I will touch but lightly on "continuous dilatation," or the tying in of the instrument. There is a patient upstairs who is now undergoing it successfully. You have tried, we will suppose, the simple dilatation, and have not made the amount of progress desired; or, perhaps, the patient's avocations may make it necessary to have a more speedy cure. In either case you may say, "If you can give me ten or fourteen days in your room, not necessarily in bed, but on the sofa quietly at home, I can almost certainly bring you from the smallest number up to the highest,"—that is, by "continuous dilatation." In "simple dilatation" the instrument is simply introduced, and at once withdrawn; in "continuous" you tie

the instrument in, and allow it to remain for several days.
You tie in a small catheter, which, if possible, is to be gum
elastic, and so that it only just enters the bladder. And you
should always take care that it is small enough to pass easily,
so that it lies loosely in the canal. Those three conditions
being granted, this is one of the safest and best modes of
treating stricture. There is a patient upstairs who has
finished the process, and to-day the house surgeon tells me
that he passed No. 11 with ease. The man has not now
the slightest pain or frequency of making water, and he has
not been so well as he is now for twenty years. He says he
is as well as ever he was in his life; he came here in an
exceedingly bad condition. He had been treated as an out-
patient, and making no progress, I advised him to come in,
and try continuous dilatation. I repeat, three conditions are
necessary to success: you must have a flexible instrument;
the point must not be far in the bladder, and it must not fill
the stricture, because, remember, it is not a mere mechanical
process; you do not want to distend the stricture as you
would a lady's glove, but you let the foreign body lie in the
passage. If you leave a No. 1 in for a sufficient length of
time, you will be able, when you take it out, to pass No. 10,
without using the intermediate numbers. You do not leave
so small an instrument there, because the stream of urine
would soon wash it out, and it is necessary to put in a larger
one to fill the canal a little more. If, when you change the
instrument, you put in one quite as large as the stricture
will admit, you produce pain and irritation, and infringe the
principle I have laid down, so that the progress is less
satisfactory. In continuous dilatation, as well as in simple
dilatation, you adopt the plan which gives the least amount
of pain and irritation to the patient, putting as little as

possible to the debit side, and as much as possible to the credit. Well, then, having ceased to tie in any longer, say at the end of six, eight, or ten days, according to the case, you will pass an instrument daily for two or three days. Then gradually increase the intervals of time, maintaining as much of the calibre gained as you fairly can. Most commonly you lose a number or two from the highest point attained by tying in: thus, if No. 12 was so reached, you will probably maintain No. 9 or 10 ; an excellent calibre for a patient who commenced with No. 1, and that perhaps not passed without much difficulty. Of course there are some few cases in which the contraction reappears rapidly, and all that has been gained by the process is rapidly lost. Clearly, some other method than dilatation will be required for such cases. This subject, which embraces the operative proceedings to be applied to the treatment of stricture, I shall speak of at our next meeting.

LECTURE III.

STRICTURE OF THE URETHRA.

You may remember, gentlemen, that at the last lecture we commenced the subject of stricture, and considered the treatment by "ordinary dilatation" and by "continuous dilatation." You will see at once that this presupposed that an instrument passed *into the bladder*. It is, of course, assumed that in treating a case by means of dilatation, whether continuous or simple, the instrument has fairly passed through the stricture, otherwise the stricture is not dilated. But all cases of stricture are not so easily disposed of. It often happens that at the first, second, or even third trial you do not get the instrument into the obstructed part, or only partially so; or it leaves the canal altogether and goes into a false passage. At all events, the instrument does not go through the stricture, and onwards, as it should do, into the bladder. Now that is a condition of difficulty which opens a new subject for us to-day. We have now to deal with more difficult cases, those in which all your care and experience, if you have any, are needed. It was said by Liston, that of all operations in surgery, there were none so difficult, none which required so much patience and care, as the passing of a catheter through an obstinate and very narrow stricture. That statement is in his last edition of the "Practical Surgery" (p. 476), and you can scarcely have a higher authority for the fact there mentioned.

Now there is one word which is often used to describe

stricture, to which I take a great objection, and I think the
sooner it is expunged from the vocabulary of surgery the
better. Such a stricture as that of which I now speak is
often said to be "impermeable." What is impermeable
stricture? Why, first, it does not denote a quality necessarily inherent in the stricture at all, but rather the quality
of the surgeon. For, you see, a stricture may be "impermeable" as regards A, but not "impermeable" as regards B,
who may pass the instrument easily enough. But, secondly,
it is a contradiction in terms. Stricture is a *narrowing* of
the canal; it is not an obliteration of it. There must be an
opening, and if there be an opening there must be room for
an instrument; it cannot be "impermeable." It is only a
question of the size of that instrument, and of skill or
patience in the management of it. The stricture always
admits urine, more or less in quantity, to pass through it,
and I maintain the truth of the axiom first enunciated by
Professor Syme, that whenever urine passes outwards through
a stricture, an instrument ought with care and perseverance
to be got in. I advise you to believe in that doctrine, not
that it is true as regards yourselves at the present time, for
I will assume that you have not sufficient experience to be
able to pass an instrument through a stricture in all cases.
If you are able to do so, I can only say that you are quite
out of place here, and need not come to learn. It is exceedingly difficult to pass an instrument in some cases, but
after a considerable amount of experience you will find that
there are very few in which it cannot be accomplished. In
the treatment of stricture, when you have really a difficult
case before you, it makes all the difference whether you act
under a belief that it is your own fault if you do not succeed,
or whether, on the other hand, you hold the dogma that

there are a certain number of cases which are "impermeable" to all surgeons. The man who holds the latter belief will be quite certain in some cases not to succeed, whereas, probably, the man who believes that in all cases an instrument may be passed with time and patience, will be very likely to succeed in all, or at all events will succeed better than the other. "Impermeable" stricture is not heard of so much now as it was twenty years ago. It was fully believed in then, and operations were frequently seen in the hospitals for it; but I will undertake to say that they are very rare now. The operation consisted in passing a large instrument down to the stricture, and opening the urethra upon it from the perineum, and then getting through the obstruction, or by it in some way, if the operator could, into the urethra beyond. It was very seldom that the true passage was followed, but somehow, by dint of cutting, a way was made for the instrument to go from the urethra before the stricture, to the urethra behind the stricture, and it was not a very successful proceeding. That was acknowledged in all books at that time as an operation for impermeable stricture, or as "perineal section." I have had occasion to perform it twice only in my life, both instances of traumatic stricture, and I believe the necessity for it to be excessively rare. I have already given Professor Syme credit for having first enunciated that doctrine, and he has stoutly maintained it, to the great advantage of patients who suffer from stricture.

But you may have complete obliteration of the urethra, which, as before said, is not stricture. This is very rare; but it occasionally happens, and chiefly after injury in the perineum, as by the breaking of a vessel, or any wound there cutting into or across the urethra. If the opening remains pervious, and gives exit to all the urine, a cicatrix occurs

STRICTURE OF THE URETHRA.

involving the anterior opening of the divided urethra, which is then closed altogether, and so the canal is obliterated in front of the fistula.

Now, how are you to deal with a really narrow and diffi-cult stricture? I will assume that you have a case in which you have tried to pass an instrument three or four times, and have failed. The first thing to do is to see the patient make water. Your failure to pass the instrument may not have arisen from narrowing of the urethra: there may have been some false passage made. It may be that there is no stricture at all. No greater failures are made than with those patients who have little or no stricture, either from the surgeon not knowing well how to pass the instrument, or from there being a false passage into which he gets, so that he does not enter the bladder at all. You first of all see the stream of water, and judge by that what size of instru-ment is to be used. And always let the instrument cor-respond with the size of the stream which you see. It should correspond in this way, however,—it should be a little smaller. You know, of course, that when a current of water passes into a narrower passage than that in which it has been flowing, the stream is more rapid than before, and when the passage becomes larger it flows slowly again ; so that the size of the stream as it flows from the orifice is not to be taken as the precise measure of the calibre of the narrowest part of the canal. You should, therefore, take an instrument somewhat smaller than the stream. There is a patient in No. 10 ward, whom some of you have seen, who does not pass a stream at all : the urine is only a succession of drops. How small, then, in such a case must be the instrument that is employed. Then there is one very im-portant thing to be considered in connection with small instruments—namely, that no more dangerous weapon can be introduced than a very small catheter, unless corresponding care, delicacy, and gentleness are exercised; indeed, it is impossible to be too careful in employing it. You see how easy it is with such an instrument as that which I hold in my hand to get into one of the lacunæ, or into any false passage, or to drive it through the soft walls of the urethra into the tissues outside : therefore it should always be used with the utmost lightness of hand. You must not hold this slender catheter tightly, resolving to get it through any obstruction ; but it must be held so lightly that if it meets any undue resistance it will slip through the fingers directly —anything rather than wound the urethra. I certainly cannot advise you to try such an instrument until you have had some fair amount of practice with a larger one. In cases of difficulty, a small gum instrument is often not of great service. I have been advocating gum elastic instruments as the rule ; but if you have to deal with a very tight stricture, and fail to pass the flexible catheter after one or two trials, you must use a small silver instrument.

Now, one word upon the subject of using force. *Under no circumstances whatever should force be used in the introduction of an instrument through a stricture or into the bladder.* That is my dictum to-day. Years ago it was matter of discussion how much force should be employed ; more years ago still you heard of people using instruments with great violence. Now I am perfectly satisfied, and I believe all modern surgeons will tell you, that no force is to be employed. It is very difficult to say what one means by force ; but what you are to understand is, that no kind of weight or pressure is to be put on the instrument which can by any possibility carry it out of the canal; and very little force will do that. The more difficult the stricture the less are you to dream of using force. You will remember that the urethra is possibly of full size up to the point of narrowing, and hence it may be very difficult to find the opening. If you use force, you will perhaps perforate the soft walls on either side, and this done, you have increased your difficulties; because, if you make a false passage, the point of the instrument is much more likely to be caught in it than to pass through the strictured part.

Remember, also, that a stricture in the course of a canal does not necessarily follow the exact direction of that canal. It may be a little tortuous; may be on this side of the axis or on that, not necessarily in the middle. You may see this sometimes in a dead body ; and you may infer it from experience on the living. When you have to deal with a very narrow stricture, take a silver instrument which you intend to guide. Do not rely upon mere groping to find the orifice. First of all, it is exceedingly dull work to be constantly groping. You should always adopt some method—any method you please, provided that it shall seem to you exhaustive of the different modes of exploring the urethra. Some of you saw the plan I adopted with a patient to-day ; and I believe that to be the best plan. You should go systematically to work, and slide your instrument from the orifice of the urethra down the one side or the other. This diagram will show you what I mean. If I go down on this side I shall probably not get into the stricture; but if I slide down gradually on the other, I shall probably get the instrument in, because you see there is less obstruction on that side. Begin, then, by the roof. The roof is the firmest part, and, by following it, will be most likely to carry the point in. The floor, on the contrary, is the softest,

loosest, and most spongy part; and will be most likely to
yield to the instrument, and give way. If your first effort
does not succeed, take the right side; if that does not do,
take the left; if that does·not do, take the floor. I know
no other method so calculated to help you through a difficult stricture. If you are very careful, you may make the
attempt in this manner for twenty or thirty minutes without
doing any damage; but if you find the patient suffering
much, if you are losing patience, give it up, or you will
probably make a false passage, and increase the difficulty
tenfold.

So much with regard to introducing an instrument into a
very narrow stricture. Now suppose you have a false passage. Of course I will assume that you have not made one
—you will be much too careful for that; you are perhaps not
the first surgeon who has seen the patient; some one else
has seen him before you, and made a false passage. There
is that case upstairs, in which the man has made a false passage himself, confessedly, notoriously, because he has taken
a large instrument and pushed it completely *into the bowel.*
He has used a No. 9 or 10 bougie, and treated himself for
stricture. He is certainly an illustration of the old adage
that " he who treats himself has a fool for his patient." He
has thrust the bougie out of the urethra through all the
tissues between it and the rectum. When he was in the out-
patient room he simply complained that when he passed the
catheter into the bladder he brought out fæces. The truth
is he never went into the bladder at all. I suspected the
cause, laid him on his back, and verified it, and, as you
know, he is now in my ward upstairs. I have made two
prolonged attempts, and I carried a No. 1 silver catheter into
the bladder to-day. You can easily conceive how difficult

that false passage must make it to get the instrument in. It is only to be done by being very careful in avoiding the side on which the false passage is. A false passage commences usually on the floor, and, no doubt for the reason mentioned, the structures below being looser than those above. When you examine a patient with a false passage, you may find the instrument passing to its very end, and still no urine flows. Hence the false notion of stricture at the neck of the bladder. When the instrument has thus passed, put your finger into the rectum, and you will know instantly whether there is a false passage; for, if so, there are only the coats of the bowel, which are very thin, between your finger and the instrument, so that you feel it very distinctly. But if the instrument is in the right passage, you feel the whole thickness of the prostate, not always very considerable, between it and your finger, still quite enough to show that you are in the right path. It is almost always in the bulbous portion that it leaves the urethra and gets under the prostate. What you are to do, then, is to withdraw the catheter two inches or so, then pass it on again, keeping as close along the upper part of the urethra as you can, ascertaining by means of the finger that the instrument is not passing into the old route. It will be very likely to do so, because it is much more easy to get into a false passage than into the right one. I have devoted as much time to this subject as I dare. I have only given you these general hints; you will arrive at the rest yourselves by practice. There is very often a case of false passage upstairs, and I wish you sometimes to verify the position of the catheter when it is in the false passage; to observe, by introducing into the rectum your finger, how very little tissue there is between it and the instrument. On the other hand, when it is properly passed, you feel the thickness of the prostate intervening. You can only appreciate this by the touch, and will learn very little more by talking of it.

One word with regard to the injection of oil. When you have a very narrow stricture to deal with, instead of oiling the instrument, it is as well to throw half an ounce or an ounce of olive oil into the urethra, holding the syringe well round the meatus. It is easy to insinuate half an ounce or an ounce of oil through a very narrow stricture. It lubricates the parts, and sometimes the urethra is distended with the oil, so that, if you can cleverly retain it with the finger and thumb, you may introduce the instrument when you have been unable in any other way. This is worth noting. It does not do well when there is much bleeding, or the tissues are torn, but when it is not so the plan is sometimes useful.

Now suppose that, following these hints you have carried the instrument through the stricture; there is a sort of grasp which is quite unmistakable. That is a sensation which you are always content to have, because, feeling the catheter "held" by the stricture, you know you are through it. But that very "grasp," which you are so satisfied to have, makes it less easy to manipulate the point of the catheter after it is through the stricture, and you have sources of danger to encounter in the urethra beyond. Thus, the mucous membrane being often reticulated with dilated lacunæ, it is easy to engage the point of the instrument in one of those, and make a false passage. Be careful never to use force in this stage, and, even after you are through, to get gently and quietly into the bladder. Here is a drawing taken from a case which exactly illustrates this condition. (Fig. 5.)

Let us pursue the case one step further. Suppose you

have got your catheter in at last, after much difficulty. Do not pull it out. You will say: "I had trouble enough to get it in; and now I will tie it in at once." It is generally safe to do that, although it is a metal instrument; and you may keep it forty-eight or seventy-two hours before removing it. Do not then be in a hurry to take it out, if the patient is tolerably comfortable. You will be excessively disappointed to be obliged to repeat your task; and it will be very disagreeable to the patient. Keep it in about three days, and then you will mostly be able to change it for a gum-elastic instrument. You then go on with "continuous" dilatation, as already described, increasing the size of the gum catheter from time to time. You get up, perhaps, to No. 10, and tell the patient, "This is a great point gained;"

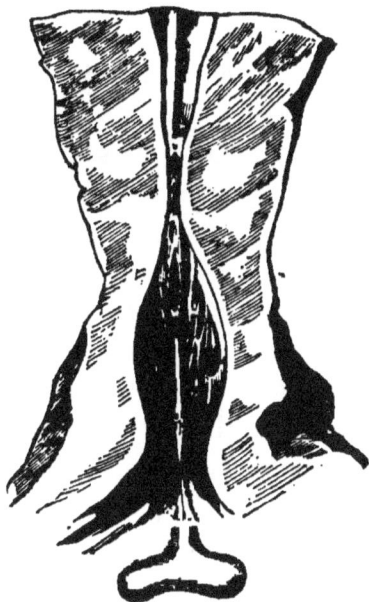

FIG. 5.—Section of urethra, showing very narrow stricture, and dilated and reticulated membranous and prostatic portions behind it.

but in ten days or so, to your great disappointment, it may be that the stricture will admit nothing larger than No. 2 or No. 3. Now it is clear that you have to do with what is called a "resilient" or "contractile" stricture. It is not a question of calibre so much as a question of contractility. The narrowing has gone back almost to that condition in

which it was at first. You can now only get in, say No. 2.
It is useless again to attempt dilatation. This is precisely
that kind of exceptional case for which you reserve other
means of treatment. Understand that you may have a very
narrow stricture, and be able to dilate it, and the dilatation
shall be fairly permanent. On the other hand, you may have
a stricture which will admit even No. 5 or No. 6 catheter
easily, yet the man will scarcely make a drop of water, and
you cannot dilate the stricture more than a number or two, do
what you will. We had an instance of that in the ward the
other day. No. 6 instrument was passed; but the man
could make no water until I had operated. These contractile
strictures have been the plague of surgeons from time imme-
morial. If you go back to old records of surgery—how far
back?—some hundreds of years—you will find that these
cases have exhausted the wits of surgeons to the present day.
All kinds of things have been used in order to overcome the
difficulty. I cannot tell you one half of the matters of various kinds which have been put into the human urethra, for
the purpose of curing it. I suppose the human stomach has
been made to receive more abominable things than any other
receptacle in or out of the human body. But if you con-
sult the old surgical authors, or even some modern books,
you will see that the urethra has been used nearly as roughly,
and that is saying a good deal. Verdigris, savin, the salts of
all sorts of metals, everything that could irritate, or that
could be imagined to be disagreeable, has been employed to
cure these unfortunate strictures. At the present moment I
need not tell you that some surgeons have employed nitrate
of silver and caustic potash—not at all mild remedies, either
of them. Now the whole question of chemical irritants, as
applied to stricture, I shall dismiss with the following words:

I believe them to be unnecessary, undesirable, and often injurious. Most modern surgeons, both in this country and abroad, have pronounced against the use of caustics and chemical irritants in the treatment of stricture. I am bound to tell you they have still some advocates: what system has not? I shall not pursue that question further.

Then what have we left? Several mechanical modes: we may rupture, or over-distend, or cut these unyielding and contractile fibres, which constitute the stricture. Urethrotomy, as it is called—division of the strictured urethra by some form of knife—is, perhaps, all things considered, the most universally employed in such cases. Now there are two kinds of urethrotomy—external urethrotomy and internal urethrotomy: the external applied from the perineum; the internal by means of the knife, or some other instrument applied within the urethra.

I speak first of internal urethrotomy. There are two modes of doing it. You may cut the stricture from before backwards, or from behind forwards, the latter being the method most commonly adopted.

First, suppose you have a stricture at or near the orifice —a very undilatable part. You get in an instrument like that which I hold in my hand—a small bistourie cachée;

Fig. 6.—A small Bistourie Cachée.

project the blade and draw it outwards, dividing the obstruction. Nothing can be more simple. By means of a screw, you can raise the blade to a small or to a large extent. It

E

should be used so as to make a rather free incision, which is perfectly safe. As a rule, the nearer the stricture is situated to the orifice of the urethra, the more necessary is it to cut, and the safer is it to do so. The further the stricture is from the external meatus, as a rule, the less necessary it is to cut, and there is rather more risk in doing so. All strictures at the meatus, and within, say about three inches of it, which have resisted dilatation—and these generally do so—should be divided. You can dilate them, but the effect is temporary; and you can divide them with ease and perfect safety. This drawing shows the condition in question. The strictures are laid open after examination, and pre the appearance of much larger calibre than really existed during life. (*See* Fig. 7.)

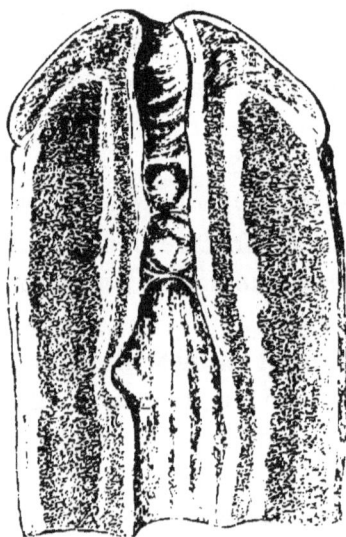

Fig. 7.—Strictures near to the orifice of the urethra.

Secondly, with regard to strictures at a distance of five inches, the reason why it is somewhat less safe to cut them is on account of the large amount of erectile tissue at that point. In my own experience of sixteen or seventeen years of these operations, I have never lost a case by either internal or external urethrotomy, either in the hospital or out of it, and I have had a great many cases; I cannot say how many—perhaps a hundred. Sometimes there may be a dangerous amount of fever, and, rarely, cystitis; much depends, I need not say, on the way in which the thing is done. External urethrotomy is usually known as Syme's operation. It is necessary always, in that case, to pass a small grooved staff or director through the stricture, to cut down upon the perineum, as in median lithotomy, and divide the stricture completely by cutting freely into the groove of the staff. That is very seldom done now. It was much done twelve or fifteen years ago ; but other means have superseded it. Occasionally the operation is still performed, chiefly when there are perineal fistulæ which require to be laid open. After doing it, a catheter is tied in for forty-eight hours, and an instrument to maintain a good calibre must be passed from time to time afterwards. Now, for internal urethrotomy in these cases of stricture in the bulbous urethra, this instrument of Civiale's which I hold in my hand is a very good one. I used it for a case here about a month ago, and the result was simply perfect. It has a small bulb at the end, concealing a small sharp blade within, which you pass through the stricture. You can then, by means of the bulb, feel exactly the position of the stricture. Having done this, you unsheath the blade about half an inch beyond the stricture, to the extent required, and draw it forward, dividing the contracted part freely. In that case, also, you should tie in a

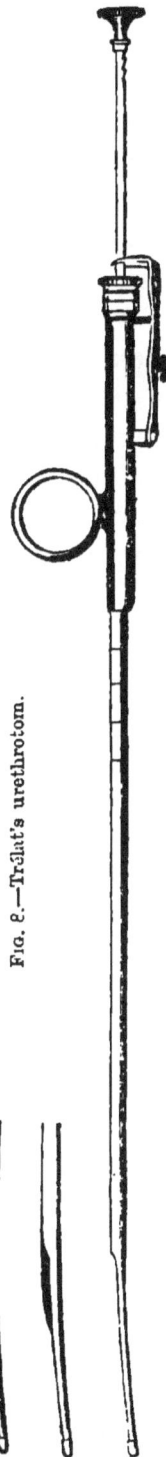

Fig. 6.—Trélat's urethrotom.

E 2

catheter forty-eight hours, and after that pass an instrument
every third day, then every week, and so on. Here are
instruments for cutting from before backwards, and of
various kinds; but I prefer the others as a rule. Here is
another which will act in both ways—a very complete
instrument of its kind, which bears the name of M. Trélat.

We next come to the mode of rupture; and here I will
show you an instrument which bears the name of Mr. Holt,
of the Westminster Hospital. He has brought the instrument into notice, and the mode of using it is his own. It
was invented by M. Perrève, of Paris, more than twenty
years ago. He used it mainly, but not altogether, for
simple dilatation; Mr. Holt uses it otherwise. He carries
it through the stricture, and then, instead of passing at
different times a succession of tubes of gradually increasing
calibre, he takes the largest tube at once, and forces it
down the urethra along the central guide, so as at one
stroke to split everything that happens to obstruct the
passage of the instrument. Mr. Holt does not tie in an
instrument afterwards. When the operation was first introduced, I was strongly repelled by the violence of the proceeding; but I examined some of Mr. Holt's cases with
him at the Westminster Hospital—now some ten years ago
—and was surprised to find how few bad results followed.
Hence I tried the plan, and use it occasionally for a urethra
which has some two or three contracted points—a condition
rarely met with—so as to deal with them all with certainty.
Rupture, however, is too rough a mode, in my opinion, for
most cases; and I prefer to carry a fine blade carefully
through the fibres of the stricture, and believe it to be the
best, and the most enduring in results. But the popular senti-
ment about a knife cannot be ignored. The British public is

not partial to a sharp edge, and is glad of almost any substitute—a feeling one can quite understand. So that it is
not always possible to do the best thing, and we have to
select that which is next best. You will then, probably, find
the proceeding by rupture a useful one when dilatation has
failed. That operation suggested to me some years ago a
different method—namely, one which I have called "over-
distending" the stricture. It is simply this. Here are two
blades, as in the instrument used by Mr. Holt, but these two
blades can be separated for a considerable interval at one
point, and as slowly as you please. In practice I do it very
slowly, so as to rupture as little as possible, and to over-
distend as much as possible, the structures which form the
obstruction. There is this fact also to be noted : I wish
you to remember that the bulbous part of the urethra—the
usual position of stricture—is also the most distensible part
in the natural condition of the canal. Supposing the external meatus to be about No. 12 (English scale) in calibre,
the bulbous urethra admits at least No. 20 or 24. Hence
it follows that no kind of dilatation or operation which is
limited in extent by the size of the external meatus, more
than half restores the urethra which has a stricture at the
bulb. It is on this account that I distend the contracted
part to at least this, or even to a larger size, or rupture it if
I please, by means of the instrument in question. I have
now used it many times, and it is certainly attended with
good results. It is to be used only for strictures within the
bulb. I have heard that it has been employed for those
within three inches of the meatus : it is a mistake to use it
for such, which, as I said before, ought to be cut. This
operation requires more care than the other. Mr. Holt's
operation needs no care after the instrument is once in place;

it is extremely easy to use it. If you once get the instrument in correctly, a single impulse of the hand forces the tube through ; and it is certainly tempting on that account. And for a considerable time afterwards you have good results from these operations. I think, however, that internal urethrotomy gives results which are more enduring than any other ; but it is infinitely more difficult to perform properly, and, without doubt, it requires a practised hand.

One thing is very necessary before you cut a stricture, namely, that you have an exact idea of its situation and extent. Also, whether there is any other narrowed spot in the canal besides the principal contraction. If No. 11 or 12 pass easily along the urethra for 5 or 5½ inches, rely on it you have only to do with a single stricture. Not very unfrequently you will find a second near to the external meatus, or at a distance of about three inches from it. Supposing that your large bougie stops at either place, and that a No. 7 or 8 only will pass through, you may, if you please, take a bougie with a bulb at the extremity and a narrow neck behind it ; such a bulb as will just go tightly through the narrowed part ; having passed it, you feel it free beyond, and, on withdrawing, it marks again the situation and length of the stricture, the latter usually very inconsiderably, and so renders it very evident. With these bulbous bougies, of various sizes, you may determine the existence of any narrowing in any part of the urethra with great accuracy, and for many years I have never operated without employing them carefully beforehand.

Briefly let me, at the close of this subject, remind you that in all cases of impeded micturition, attention to the general health often aids in a considerable degree to mitigate the local troubles. Do not overlook the state of the digestion. If this is unsatisfactory, if the bowels are unduly constipated, the troubles of the bladder and urethra will be much increased; and frequently it happens that a mild mercurial, followed by a dose of Glauber's salts, or of Friedrichshall water, in the morning, gently unloads the liver and bowels, and greatly relieves the most distressing symptoms. Then take care of what your patient eats, and more especially let his alcoholic drinks be in moderate quantity, and of the mildest kind.

There is only one other word I have to say: do not be partisans of any single method. You hear one surgeon say, " I always follow such and such a method: there is nothing like it." Or another, that he always adopts the proceeding of M. Civiale; and a third that of M. Maisonneuve, and so on. There has been great fertility in inventions of this description, especially in Paris, and you may see from several of them very excellent results. Do not confine your selection to any one method, whether I or any one else recommend it; but have every resource at your disposal. If you have much to do with stricture, or with such complaints, I can only tell you, you will want all the resources within your reach. Consider them carefully, and select for each individual case that method which appears in your judgment to be best adapted for it.

LECTURE IV.

HYPERTROPHY OF THE PROSTATE AND ITS CONSEQUENCES.

GENTLEMEN,—We may pass from the subject of stricture to another very important complaint, and one of common occurrence—viz. hypertrophy of the prostate. It is one which affects a large number of elderly people, and thus the practitioner is almost certain to come in contact with it pretty frequently. Hence the necessity for our studying these cases closely, and the more so because we do not see them very often in the beds here, most of them being treated as out-patients.

It was formerly stated, on the high authority of Sir Benjamin Brodie, that " when the hair becomes grey and scanty . . . . . . the prostate gland usually, I might perhaps say invariably, becomes increased in size ; " and that is the impression which a large portion of the profession have respecting it. It was certainly that which was generally received when I first began to make some special researches in reference to this matter, now some ten or twelve years ago.   I was then at the pains to examine after death all the bodies of male patients over fifty-five years of age who died in the Marylebone Infirmary ; and afterwards, in Greenwich Hospital, the inquiry was pursued by Dr. Messer and myself. I took care to dissect each prostate very carefully, and dis-covered that, so far from the presence of prostatic enlarge-ment being the rule, this condition was quite exceptional. I examined about two hundred cases—not picked cases, but all who died consecutively within a certain period, and I found that about one in three exhibited after death some enlargement of the prostate. But of those do not suppose that anything like a large proportion manifested symptoms; for only about one in seven, not more, had symptoms of the complaint. So that you see it is not to more than one in (let us say) ten men who live beyond fifty-five years of age to whom you may expect to be called to afford any relief for this affection. That, no doubt, is a large number. If you suppose that one man in every ten approaching sixty years of age has symptoms of enlarged prostate, you will see at once how often, if you have anything like a large practice, you may be called to attend to these cases.

We will now consider one or two anatomical points connected with enlarged prostate. This organ is, as you know, composed of two lobes and a median portion. Now, the part affected with hypertrophy very much influences the results in relation to the function of micturition. It is not necessary that there should be much enlargement of the prostate in order to produce very severe symptoms. On the other hand, you may have a very large prostate, and may have almost no symptoms. Almost the largest I ever saw, as big as a small cocoanut, produced very little obstruction to the flow of the urine. Thus, if the median portion of the prostate is only slightly enlarged, there may be retention. Let this figure [diagram] represent the two lobes, and this the median portion. If there is a small nipple-like projection at the median portion, just filling the internal orifice of the urethra, that may be quite sufficient to prevent every drop of urine passing by the natural efforts. Sometimes there is a considerable enlargement on one side, so that the passage is circuitous; and you will sometimes find

the catheter carried to the right or left, according as the
prostate may be large on one side or the other.   I show you
several examples: these are depicted at Figs. 9, 10, 11.

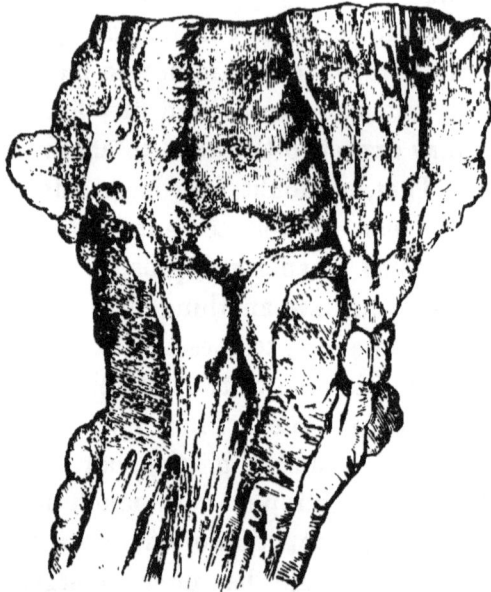

Fig. 9.   Section of bladder and prostate, showing marked but not great enlargement of lateral lobes and median portion.

You will remember, then, that if you have occasion to
examine a patient, and you find a very large prostate, it
does not necessarily follow that he should have great difficulty in passing his water; and, on the other hand, although
you may find inappreciable enlargement by rectal examination or otherwise, you may not therefore conclude that all
his troubles—and they may be considerable—are not due to
this complaint.

I will now say a word as to the age of the patients.   I
never saw an enlarged prostate (I mean, of course, hypertrophy, not enlargement from inflammation and other causes)

before the age of fifty-four; and if I have not seen such a case, you may say that it never or very rarely occurs. The usual time at which it begins to show itself is from fifty-seven

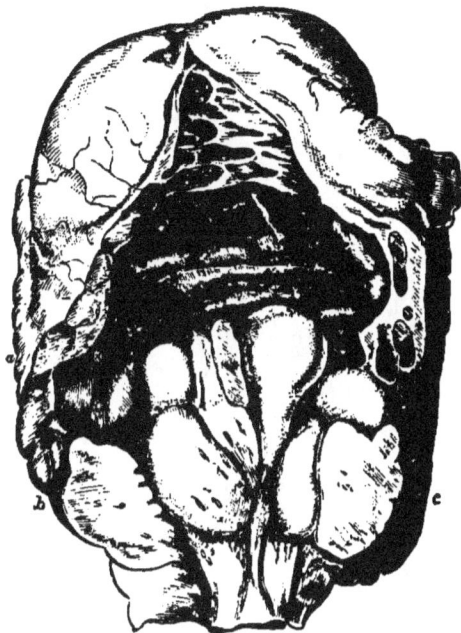

Fig. 10.—Section of bladder and prostate, the former hypertrophied, the latter forming prominent tumours within the bladder.

to sixty. If a man has it at all he will have it generally by sixty. If he gets over sixty-five or seventy, he may have it, but in a less degree; that, however, does not commonly happen. I have examined the bodies of men at ninety, without the slightest enlargement. You see, then, that it is by no means necessarily connected with age. The man who escapes it at sixty-five will be very likely to escape it altogether, or nearly so.

I speak next of the symptoms. An elderly man comes to you and says that recently his water has not passed so easily;

that it has issued in a dribbling stream, and he cannot propel it; that he requires to micturate a little more frequently,
especially in the morning—probably two or three times

FIG. 11.—Section of bladder and prostate, the latter forming an enormous
tumour. A lateral view.

while he is dressing, after which it becomes less troublesome ; but at night it is rather more so than during the day.
Then if he does not say much about pain—which, of course
will excite your suspicion of calculus, or some other complaint—you will say, " This is probably a case of enlarged
prostate."   You do not necessarily proceed at once to pass
the catheter, but you will ask the four questions already referred to.   You will ask how frequently he makes water, and
then, in connection with frequency, whether the water ever
passes without his being aware of it, or without his willing
it.   In many advanced cases, you will find that some urine

passes during a violent effort, such as coughing, or at night, during sleep. If so, it is probably a case of rather long standing. But if the patient only speaks of the frequency of making water—not inordinate, but moderate frequency—then you ask for pain, and if any, whether before, during, or after the passing of water. If before, and the patient is relieved by making water, it is probably hypertrophied pros-tate. If it is after, you may expect calculus, which comes into contact with the mucous membrane of the bladder when the water is gone; whereas, if he has a distended blad-der, as he is likely to have, with enlarged prostate, it is pain-ful as distension occurs, and becomes less so as the urine passes off. Then you inquire about the character of the secretion, whether it is clear or cloudy. In most cases, at the commencement, it is clear. In a great number of cases of prostatic enlargement, although the bladder has not been emptied for months or a year, the water is still clear. On the other hand, if it is an advanced case, the water will cer-tainly be cloudy. And while you are talking of that you will also ask about the stream. You will generally find that it is a dribbling stream, different from the stream in cases of stricture. In stricture the stream is often propelled exceed-ingly well, although it is no larger than a thread; and so long as there is a stream, the patient can act upon it by will, so as to make it stronger; whereas, strain as he may, in pros-tatic enlargement, he often cannot influence the stream, ex-cept for the worse. It may happen, from the median portion being forced by the straining into the passage, that the patient voids urine less well the more he strains. Generally speaking, the expelling apparatus at the neck of the bladder is involved in the enlarged prostate, and ceases to act; so that with all his straining he cannot make much difference, and the stream is not propelled with any force. Then you ask the fourth question: "Do you pass blood?" Usually, in the early stages, the reply will be in the negative, although a little may appear after much exercise, so far suggesting stone. You ask the patient to let you see him pass water, if he can, for that will help you materially. If the stream is such as I have described, you will infer that it is a case of prostatic enlargement.

Then you complete your diagnosis by mechanical means, and for this purpose you will first use a catheter. You should invariably make the patient pass water before you begin, because your object is not merely to ascertain whether enlarged prostate exists, but, what is much more important, what the effect of it is, how far it is a barrier to the exit of urine from the bladder. The great point to him and to you is not merely the size, or condition, or shape of the prostate, but to what extent is it a barrier to the exit of urine. It is the quantity of urine left behind which will determine future treatment. I should advise you to use a gum catheter, well curved, and certainly not a small one. As in stricture, always begin with a catheter of not less than 8 or 9, and of course without a stylet. In passing it, keep the shaft well back in the groin, so as to maintain the curve. As soon as you have arrived at the bladder, carefully empty it, and note the quantity withdrawn. It may vary greatly, from an ounce up to almost anything you please. I have drawn off six pints, but that is a very large amount. You may find commonly from six to twenty ounces.

Now, with regard to the employment of the instrument for patients with the symptoms described, do not forget that frequency of passing water, and still more the passing of it involuntarily, indicate the necessity for the catheter. It is remarkable how common are the errors, not merely of patients, but of practitioners, on this point. They are apt to be misled by the fact that the patient insists, " I do not make too little water; I am making water too frequently, and too much of it, and I am sure my bladder must be empty. Tell me how to retain my water, and I shall be much obliged to you. Don't think of drawing it off." It is surprising how that sometimes influences the practitioner. Nevertheless, these are the very circumstances in which you should pass the catheter and ascertain the fact. Always bear this in mind (and I wish, figuratively speaking, to render that sentence in the largest capitals), that INVOLUNTARY MICTU- RITION INDICATES RETENTION, AND NOT INCONTINENCE. There are a few exceptions to the rule, but very few. Most of the mistakes that are made on this point arise from the use, or, as I shall show you, the abuse, of the word "incon- tinence," which means, of course, that the bladder is empty; and certainly, when the bladder cannot hold its contents, its condition is rightly described by the word incontinence. Now, that happens only in very uncommon but well-defined circumstances, such as in some cases of cerebral or cerebro- spinal paralysis, and in rare injuries to the neck of the blad- der; and in these the urine runs off as fast as it comes from the ureters, the bladder having ceased to act as a reservoir. You see this one external physical sign is the same in these cases and in those in which the bladder is distended with urine; that is, there is urine dribbling off by the urethra. But mark how totally different are the two conditions in question : in one the bladder is full, in the other the bladder is empty. Whenever, then, you meet with this involuntary flow of urine miscalled " incontinence," do not confound it with the condition in which the bladder is empty. Rely upon it the bladder is full, and the only way of relieving the patien is by the use of the catheter. I lay great stress upon this, because I have seen lives sacrificed to a forgetfulness of this point. I have made post-mortem examinations of persons who have died from the effects of retention undiscovered during life, misunderstood because the urine constantly passed off, as it was supposed, " so freely."

Now we know that our views of things, and our consequent acts, are very much determined by the manner in which we use and apply words respecting them, and it is impossible to be too clear and defined in all our language, especially in that which relates to pathological conditions and surgical practice. I cannot express to you how strong my sense is of the importance of this matter; hence I have made it my constant business to point out the common misuse of terms in connection with this subject.

First, then, the term incontinence, which means the bladder is empty, or " cannot contain," should never be employed by you to denote the phenomenon that the patient's urine flows involuntarily ; for, as we have seen, in that condition the bladder is generally full. It is better to speak of it as " involuntary micturition," without reference to the cause, and when this is found to be distended bladder, to use the term " overflow." Then, remembering always my maxim, that " involuntary micturition indicates mostly retention, not incontinence," you will never make the fatal blunder I have spoken of, and which I assert to be so common. This, too, assimilates our usage very nearly to that of French surgeons. The French, with their more logical use of language, speak of the bladder as " engorged " and "overflowing," but never as " incontinent," except to denote that rare condition in which the bladder is always perfectly empty. I have, therefore, long been in the habit of denoting a bladder which is full, but allows surplus urine to run off little by little against the will

of the patient as an "engorged" bladder, and the phenomenon thus described as "overflow;" and I hope you will
do so too.   This brings us, by the way, to another mis-
application of terms.   In this country, the condition of the
organ just alluded to is often called "paralysis" of the
bladder, and the unfortunate word leads to mistakes in
practice.   The bladder is rarely paralyzed.   I know nothing
of it except as an effect of spinal or cerebral changes.   The
bladder is never by itself the subject of paralysis, meaning,
of course, an affection of the nerves, either central or
peripheral.   It may be unable to expel its contents, because
there is mechanical obstruction, as enlarged prostate, stricture, or impacted stone, or because the muscles have lost
their power of contracting from long over-distension ; but
this latter is "atony."   This inability is in neither case due
to impaired nervous supply—a subject to be considered at
some subsequent meeting.   [*Vide* Lecture XI.]

After this digression, which its importance must excuse,
we will go on to complete the diagnosis.   While the patient
lies on his back, you place your finger in the rectum, and
examine the size of the prostate, whether it is very tender,
and whether the enlargement is more on the right or the
left side.   Of course you do this as gently as you can.
The finger should be well oiled and very slowly introduced.
The best position for the patient is lying on the back,
because you can press the other hand above the pubes, and
gentle pressure there brings the bladder and prostate near
the finger, and you can ascertain whether the bladder is distended or not.   These are the points of diagnosis which it
is desirable to ascertain, and beyond these it is not usual or
desirable to carry your inquiry.

We shall now come to the treatment.   The *medicinal*

F

treatment of enlarged prostate may be dismissed in a few
words. There is nothing to be done for it—that is, you
cannot diminish the hypertrophy. There is often temporary
enlargement from congestion; and that you can do something for. But hypertrophy cannot be diminished by any
known means. All sorts of things have been tried : iodine
in all its forms. And there is scarcely anything in the
Pharmacopœia that presented a chance of doing good that
has not been tried. We must simply say that, for the
present, we know no means of combating the enlargement
itself. But much may be done by way of palliative to
the results of the complaint; and much of this treatment
is mechanical. This will consist first in relieving the partial retention, which has become habitual, by the catheter.
I have reserved for this place what I have to say about the
instrument itself. The reason for preferring, as a rule, the
gum catheter for this purpose is as follows :—Different
curves are required for different patients, and the English gum
catheter I will back against any other instrument, English
or foreign, for general use. Unlike the French catheter,
which is admirable for softness and elasticity, you have
the power of making it assume any form you like—a quality,
perhaps, not so often rendered available as it might be, but
which to my mind is of immense service. The instrument-
maker generally curves the catheter pretty much in this way
(see Fig. 12) : the point straight and not well curved—the
worst form in which you can put a catheter for use. For
prostatic enlargement you require a catheter well curved *to
its very point*. You should keep the instrument on an over-
curved stylet for a month or so before employing it; and you
will then find it easily assume the proper form, when you
will pass it, as I need not say, without a stylet. If you re-

quire a stiff instrument, a silver catheter should be selected ; not a gum catheter with a stylet in it. To return : you want the point, of course, to be carried over the obstruction

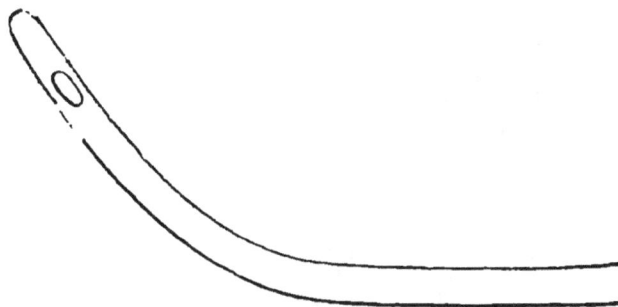

Fig. 12.

formed by the enlarged prostate; and as the heat of the urethra always relaxes the curve, whatever it may be, by the time it arrives at the neck of the bladder, the ordinary gum catheter, as you have it from the maker, becomes nearly straight, and will not pass over the enlargement. [Diagram.] Now, when you have a catheter which has been well over-curved for a month or two, you remove the stylet and turn back the shaft, so as to undo the extreme curve and produce an ordinary one. And what happens when you pass it? In spite of the heat of the urethra, the catheter has a tendency to curve more, instead of less, as it passes down the passage. And this is precisely the difference between success and non-success. That little manœuvre I regard as of extreme value. It is a very simple thing: keep the catheter over-curved—not for stricture, but for enlarged prostate ; then turn back the shaft immediately before using. The curve gradually increases as it goes onwards, and it passes over the enlargement into the bladder. This is so simple that it seems

F 2

scarcely worth making so much of ; but I can only tell you
that I know nothing of its kind that exceeds it in value.

Then there is another thing : you may want a special
curve for a particular case. We have silver catheters with

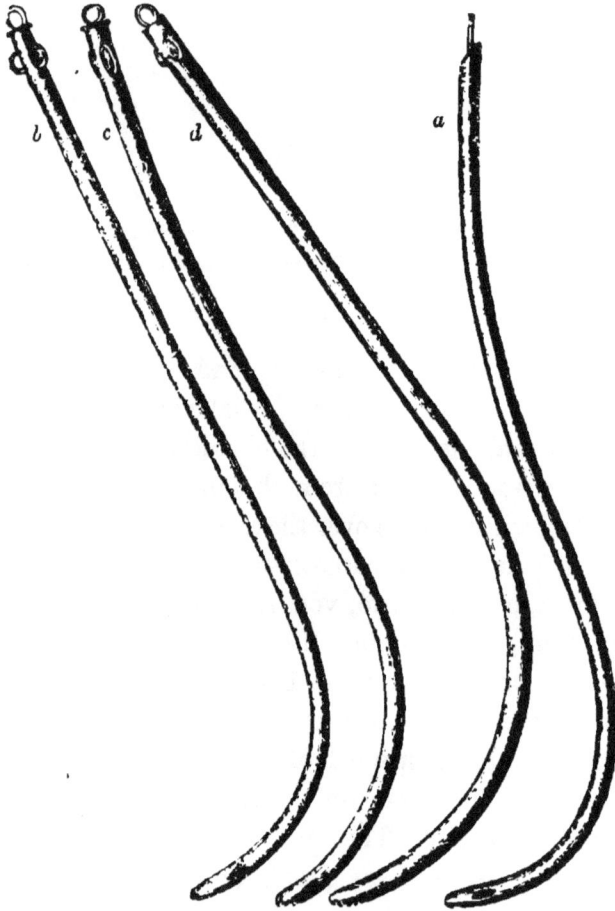

Fig. 13.—*a*, Gum catheter mounted on a stylet of the proper curve for use ;
b, *c*, *d*, silver prostatic catheters of different curves.

various curves (*see* Fig. 13). Here are several which are
very useful. But the English gum catheter possesses a

quality which, as I have before told you, is not found in
any other: put it into warm water and bend it into any
form you like; them dip it immediately into cold water, and
it will maintain the required form. But the best form so
produced may easily be spoiled by your mode of using it.
Of course the curve must not be altered while the instrument is passing through the anterior part of the canal, for
it is at the posterior part that this form is required; the
shaft of the catheter must be kept closely back in the groin,
and the penis brought round the curve, so as to preserve
the latter until it reaches the deep urethra, when, by well
depressing the shaft, the point will rise over any obstruction
into the bladder.

The general treatment is not to be disregarded; and here
I shall defer a good deal of what I might otherwise say
until I come to speak of chronic cystitis, which will occupy
a subsequent lecture. Cystitis is associated with so many
diseases of the urinary organs, that I may as well refer to
the treatment under that head, instead of taking it separately,
and recapitulating it in connection with each disease. But,
as far as the general treatment of prostatic enlargement goes,
the main thing is to prevent local congestion. You must
tell the patient above all things to avoid anything like chills
affecting the pelvic region, sitting on cold seats, exposure
to cold; too much excitement, sexual or otherwise; long
journeys, riding in jolting carriages,—all of which tend to
produce pelvic congestion, and to interfere materially with
the condition of the prostate; since it very readily becomes
temporarily enlarged from these causes, and most of the
troubles which the patient experiences depend upon them.

One other point has relation to the action of the bowels.
You may make a man with enlarged prostate very comfort-

able if you keep the bowels in gentle action. If he has constipation, and scybala lodge in the rectum, their presence is often the source of great discomfort. Sometimes a simple enema of warm water gives instant relief; but, if necessary, the action of the bowels must be provided for by means of gentle laxatives, such as senna, manna, bitartrate of potash, sulphur, or by sulphate of soda; anything which will act mildly, quickly, and without irritation, will keep him in a very different condition from that which is associated with habitual constipation.

I will devote the few minutes that remain to those cases of prostatic enlargement in which the difficulty of passing the catheter is considerable, and there is retention of urine. You may have a patient in whom prostatic enlargement has made itself manifested rather suddenly; he may have had symptoms which he has not noticed, but he is suddenly attacked with congestion, he cannot make water, and he is in great distress in consequence. It is not a question how long a time is to elapse before the catheter is to be passed; you should relieve him at once. You find there is a distended bladder, evidenced by dulness of percussion above the pubes. Possibly, before you, others may have tried, and you may be called to pass the catheter in circumstances of some danger to life. Now, in the first place, you should be careful with regard to position. I advise you to pass the instrument with the patient in the lying position first, if the bladder is very large; if it is not very large, perhaps it is as well, or better, to pass it standing. You can empty the bladder better in the standing position; but if you find that the bladder is very large, make the patient lie down before you draw off the water. I have known great mischief arise from drawing off a large quantity of water from a patient when in a standing position. I have even known death occur suddenly from this cause. If I had time, I could tell you of a case in which a charge of manslaughter was brought against a surgeon in a court of justice in relation to such an occurrence. The circumstances were all well known to me, for I was there to defend a brother practitioner, who was unjustly charged. In that case a catheter had been passed in the upright position, and the patient fell dead from syncope, when six pints of urine had passed; just as a patient with ascites might do if you tapped the abdomen in that same position. No doubt it was an error, but nothing could be more monstrous than to make it the ground of a criminal action. It is a very instructive case, and I mention it to show that in cases where the bladder is large, especially in old men, you may have fatal syncope, arising in the way I have described. I always take care, if I find the bladder is large, to pass the catheter when the patient is lying down. Take care also, in these circumstances, to draw off only part of the urine; and after a quart or so has passed, wait a little before you empty the bladder. If you fail to pass a gum-elastic instrument, by all means use a silver one, especially if false passages have been made. The silver prostatic catheter is sometimes essential; that is, one which is much longer and has a larger curve than the ordinary instrument. They are generally made longer than most cases require. Indeed, the common silver catheter, about No. 10, or, at most, one which is only two inches longer, suffices for the majority of cases. Very rarely is the fourteen-inch catheter necessary. Sometimes, when these fail, a catheter with a short beak, like the sound, or lithotrite, will pass easily. And always remember that anything like obstruction at the end can only denote that you are out of your path. No force should be employed. It is not a narrow passage; it may be a little close, perhaps, when you reach the prostate; but if you find any obstruction, you should withdraw and find another route, to the right or to the left. Here, also, let me repeat, never use force under any circumstances whatever.

It may be said, "Why have you not mentioned hot baths and opium?" In cases of retention from prostatic enlargement, there is a serious objection to the expectant method; you must take into account the future condition of the bladder. If you permit the bladder to remain over-distended, say for a day or two, the fear is that it will not readily contract again. In the case of an old man's bladder, thoroughly distended by long retention, it is very likely not to recover its powers. Although the patient has made water fairly up to the time of retention, if you leave him too long with opium, hot baths, and treatment of that kind, the bladder is getting bigger all the time, and you will very likely have more trouble with it afterwards than if you had relieved him with the instrument at once.

Again, if you have had a great deal of trouble in passing the catheter, I suppose you must leave the instrument in; but it is not good practice in these cases. Rather withdraw it, and use it again; because the prostate will be injured by an inlying catheter. Unlike stricture, which is often well treated by an inlying catheter, the prostate is always more or less irritated by it; but less harm accrues from a flexible than from a silver instrument. Better still if you can pass a vulcanized india-rubber catheter, which is the only one which lies harmlessly in the urethra in a case of prostatic enlargement. It may sometimes be insinuated by a series of short quick pushes, thus [illustrating the method]; or,

failing in that manner, by mounting it on a stylet of any
curve which may be desired, and subsequently withdrawing
the stylet. Nothing is easier than to keep it in its place,
and it has the merit of rarely becoming encrusted with phosphates; while it does not necessarily prevent the patient
moving about his room. In short, the vulcanized catheter
is sometimes of the greatest service.

Subsequently, as the canal recovers, should the bladder
not regain its tone, and it is therefore necessary to use the
catheter two or three times in the twenty-four hours, you
will probably in the course of time teach the patient to
relieve himself; and he will often, with a little tuition,
succeed remarkably well. With respect to the frequency
with which he should be advised to do this, it will depend
mainly upon the amount of urine left in the bladder after
a natural effort of micturition. The patient's own feelings
will aid in answering the question; but if you find him,
after having passed water, retaining say six ounces, a
catheter passed every night and morning will probably be
sufficient. The bladder must be emptied, or the urine will
in time decompose, become irritating, and chronic cystitis
will result. If a pint is left behind after making water, the
bladder should be relieved, as a rule, three times a day. If
the patient makes no water by his own effort, he should be
relieved four or five times, or even more—certainly not less,
as a rule, than three or four times in the twenty-four hours.

You often find (and this is a matter of considerable importance) that although up to the time of the attack, or of
the necessity for the use of the catheter, the water has been
perfectly clear, yet, after you begin to use the catheter
habitually, the patient gets more or less cystitis, and is
feverish and unwell. That stage has very often to be passed
by those persons who suddenly change from a natural mode

of micturition to an artificial one. It requires some judgment to say when that change should be made ; but as soon
as you find it necessary to pass the catheter regularly, the
patient will often show some change in his general condition;
and you should be aware of this, and watch the results. Sir
Benjamin Brodie was the first to point out, in his valuable
lectures on the Urinary Organs, that sometimes patients
gradually succumb to a low or feverish condition after
beginning to use the instrument. The remedy, or rather the
preventive method, is this :—Do not empty the bladder on
each occasion of using the catheter at first. If the patient
has been in the habit, perhaps, of retaining at least a pint
of urine after he has made water, it is a great change for
him to have the bladder quite empty two or three times a
day ; and it is thus that the organ becomes irritable, the
urine charged with pus, and then he loses his appetite,
becomes feverish, and is sometimes in danger of losing his
life. The rule under such circumstances is to proceed cautiously. Instead of drawing off a pint, draw off half a
pint; leave some behind, and so make a compromise
between the condition of the bladder and the general condition of the patient. Draw off half or only a third of a
pint ; you will thus relieve him a little, and so gradually,
in the course of a month, you may accomplish the entire
emptying of the bladder, and all will go on smoothly and
well. Notwithstanding all your care, now and then you
will find a case in which during this process the tongue
grows slowly more red, dry, and contracted ; the powers of
life gradually fail ; the senses become impaired, and the
patient sinks. You will always find in such cases, by
autopsy, old-standing pyelitis, with dilatation and injury to
the renal structure, and you will know that in no circumstances could the patient have long survived.

LECTURE V.

RETENTION OF URINE.

GENTLEMEN,—Retention of urine is the principal subject for to-day; and if we at all realize what is the condition of a patient who suffers from retention, with the acute and constant pain which it involves, we shall feel how important it is to relieve him, not only as easily, but as early as possible. And there are very few cases in which you will meet with more gratitude if your treatment is skilful and prompt. For not merely are his physical sufferings extreme—and I suppose every man who has been so situated as to be unable to relieve a distended bladder, even for a short time, has had some glimpse, though only a faint one, of the distress occasioned by inability to pass water for several hours, or even days—not only, I say, is the pain intense, but there is extreme anxiety of mind also. He fears that the bladder may burst (a circumstance, however, of exceeding rarity), and he always looks forward with gloomy prospects to the consequence of not obtaining relief.

Now, while retention of urine is very common in the hospital, it is not so in private practice. The circumstances of exposure, the more hazardous callings which men have who form hospital patients, determine this difference; therefore, when met with among the higher classes, it is always a very grave matter, and in all circumstances, wherever encountered, will demand all your care and skill.

Then, again, if you are successful, the relief which you

afford is instantaneous. It is not like the questionable result of a dose of medicine, which a sceptical patient may persist in attributing solely to our great ally—the curative power of nature. There can be no uncertainty as to the result of your treatment if, after twenty-four hours of agony, relief follows your dexterous use of the catheter, and the two or three pints which the patient was unable to void are withdrawn by your hand. He tells you that he is " in heaven " —a common expression with such patients—and he never will doubt for a moment that you were the author of his " translation."

Retention comes before us in three typical forms, each requiring a different species of treatment. There may be some instances which cannot be absolutely so classed, and some the characters of which belong equally to two forms; still, for convenience, it is well to adopt this classification. First of all, you may have retention occurring in a young and healthy man who has no stricture; next, it may occur in an older man who is the subject of confirmed organic stricture; and, lastly, it may occur in a man who is neither young nor hale, and who has no stricture, but has an enlarged prostate. Of the last I have nothing to say; we have already discussed that subject, and the mode of relieving retention in connection with enlarged prostate. But I shall ask your attention to the two other conditions—namely, retention from inflammatory swelling, and retention arising from organic stricture.

With regard to the first kind, you will probably learn a history something like the following :—within a month or six weeks the patient, who is generally a young man, has had gonorrhœa. He has obtained considerable relief from treatment, and has in consequence allowed himself some

relaxation of the regimen to which he has been lately sub-
mitting. Thus, perhaps, he has indulged a little in the use
of alcoholic stimulants, has taken some unusual amount
of exercise, a game of cricket or the like, and, after being
over-heated, has been sitting on a cold stone or damp grass ;
or, lastly, he may have indulged in some strong emotional
excitement. Under those circumstances what is called
"inflammatory stricture" may be produced. Now let me
say, repeating myself slightly, that that condition has no
right to the name of stricture. First of all, the inflammation is probably at the neck of the bladder or in the prostate. It is difficult to ascertain that, except by inference,
confirmed, however, by a rectal examination ; for, happily,
we very rarely make post-mortems under such circumstances,
as the patient does not succumb to the complaint. But
what almost to a certainty takes place is a degree of inflammation and swelling of the prostate, not in the least resembling stricture—that is, it is not a circumscribed narrowing
at a particular part of the passage, but a tumefaction of the
prostate gland, which prevents the expulsive apparatus of
the bladder acting and discharging its contents. That is
usually the real condition in what is called inflammatory or
spasmodic retention.

This condition of the prostate resembles that which
affects the tonsils, and which we call inflammatory sore
throat. Both complaints consist in the enlargement of
glands which more or less surround narrow passages, and so
interferes with the functions of those passages ; both occur
rapidly, and may be produced by external cold.

Now, what are the early signs of inflammatory retention ?
First, there is usually some cessation of the gonorrhœal discharge. Just as is the case of orchitis, where the urethral
inflammation is supposed to subside and to attack one of

the testes, the inflammation of the prostate is similarly associated with diminished discharge, and if you examine by the rectum, a tender and swollen condition of the prostate will be discovered. Then the stream of urine grows smaller and smaller, and in a very short time the patient loses the power to relieve himself altogether. He is feverish, very restless, and suffers severe pain about the lower part of the abdomen and in the perineum. Those who are the subjects of stricture may have become in some measure accustomed to the difficulty, but when an active young fellow is thus attacked for the first time, he is in a state of extreme distress.

Now as to the treatment of such a case. The patient desires ardently to be relieved immediately, and declares he cannot endure his sufferings. You see him bent nearly double in order to relieve the pressure of the abdominal muscles on the bladder, and he is even breathing shortly and quickly to avoid their action there. The old treatment in such cases—the classical treatment of twenty years ago— was bleeding from the arm or perineum, repeated hot baths, and large doses of opium, so as to enable the patient to bear the pain and dispense with the catheter. The reason assigned was, that in an inflamed state of the canal, you might do more harm than good with a catheter, and that it was therefore better to mitigate pain by the means described. I have told you that I dissent from that treatment altogether, although it is still employed to some extent. For, first, you must look at the after-consequences; and if you allow a young man to remain thirty-six or forty-eight hours with an unrelieved bladder, because you fear to use an instrument, permanent mischief may be done. I have seen patients who for years have been unable to empty the bladder after treat- ment of this kind. Extreme and continued distension of that organ sometimes destroys or permanently diminishes its contractile power, and produces a condition which is properly termed "atony of the bladder." Therefore, if you pass the catheter, even at the risk of doing a little mischief to the urethra, I am disposed to think you would be wise in incurring that little risk rather than expose the patient to the other danger. But then it ought to be done without such risk. For my own part, I always take a moderate-sized gum catheter—one not larger than No. 6, as a large one gives in these circumstances unnecessary pain—which has been tolerably well curved in the way I have before described, since it has to enter over a swollen prostate. In this manner there is generally no difficulty in relieving the patient, who is exceedingly grateful for what you have done; whereas if you put him through that long process, and he relieves himself ultimately, he thanks you for little, comparatively speaking, and he runs the risk to which I have referred. But in the event of the gum catheter not passing, you should try a silver catheter of the same size. I believe one of the first persons to denounce the old plan of bleeding and hot bathing was Mr. Guthrie. If you turn to the racy writings of that experienced and practical surgeon, you will find an anecdote in connection with this subject. He relates the account of a visit which he paid to a patient in the circumstances of retention I have described, and the reason, in strong and graphic language, why he then gave up for ever the bath and bleeding practice, and passed the catheter at once for such cases in future.

So much for the inflammatory condition of the prostate, producing obstruction to micturition. I need not refer at any length to spasmodic retention, which rarely happens. At the same time it may be as well to say that no doubt where an inflamed condition of the urethra exists, spasm of the muscles may co-exist; but the precise share which each takes in contributing to the result will not influence the treatment.

Now I come to the second form of retention—viz. that depending upon organic stricture. Here we generally have to do with an older man, because it is rare to find a young one suffering from confirmed organic stricture. As a rule, to which there are exceptions, such a patient mostly has stricture ten or twelve years before he gets complete retention. First of all you have to ascertain that it is stricture. You will probably find that he suffers less acutely than in the previous instance, but still a great deal ; the progress of the case has been more gradual, and the derangement has not necessarily been brought on by any great or sudden imprudence. He has been passing water with difficulty for weeks or months, and at length some slight thing has produced a condition of absolute retention; the last ounce has broken the camel's back. Or it may not be absolute retention as before ; there may be some dribbling, indeed the patient may have been relieving himself in that inefficient way for days, but the bladder is greatly distended, and to all intents and purposes it is a case of urgent retention. You find probably also that the patient is accustomed to instruments. Now, what you have to do is to take an instrument of middle size, and pass it down to the seat of obstruction, to see where it is. You will probably find it four or five inches from the external meatus. You should then take the finest gum catheter and endeavour to insinuate it into the bladder, and if you are sufficiently fortunate to do that, you should tie it in at once, so that you may have no further trouble with it. But that is not a very difficult case of retention. Supposing you do not get the gum catheter in, I should then recommend a small silver one, either No. 1, or even smaller, and use it in the manner I described to you in the lecture before last. Notwithstanding all your care and skill, and those, perhaps, of your friends whom you may have called in, the instrument is still not passed. There may have been false passages (they are easy to make), and there may be such difficulty that it is almost impossible for any one to pass a catheter after your failure. We then come to the question : what is to be done next ? Well then, first, much may be effected for these cases by opium and hot baths. Suppose the water is dribbling off, and you shrink from the *dernier ressort*—that is, puncturing the bladder, or other operation to relieve the retention of the urine,—a safe middle course may still exist for some of these cases. The patient up to this time may have been exposed to cold ; let him have the benefit of a warm bed and hot baths, with large doses of opium,—and you must be very liberal with opium if you use it at all,—so as materially to mitigate the involuntary straining, which he can no more help than he can help breathing, and which is utterly unsuccessful as regards the contents of the bladder, often making matters rather worse than better. The result may be that the water will dribble off more freely than before, and you may find, after two or three days, that it will come in a larger stream, and that then you can pass the instrument without much difficulty. The patient may often be saved an operation thus, if there are grounds for declining to perform it. On the whole, however, I do not advise waiting very long ; still it is better that the patient should be temporized with in that way than that he should be damaged by an unsafe hand, either with knife or catheter. Most men, indeed, are quite sufficiently confident in their powers to rely on instrumental

G

methods when they find that a patient cannot make water.
Still, if you are convinced that you are not doing any good
with the catheter—still more that you are doing mischief,
you can in most of these cases have recourse to opium or to
an inhalation of chloroform, with hot baths and fomentations, with success, as regards the immediate and urgent
condition.

But we will assume that you have done all that you can
do in this way, and that the question of relief by some other
means must be met. The bladder is increasing in size, notwithstanding your treatment. You examine the supra-pubic
region carefully, and find a tense and perhaps large tumour
there, reaching to the umbilicus, or nearly so, more like a
uterus than a bladder. In some old cases of stricture there
is not necessarily large supra-pubic dulness, for the bladder
is thickened and contracted. Introducing the finger into the
rectum, you find there also a swelling, produced by the distended bladder, and you then seek to obtain the sensation
of fluctuation. If, placing my hand above the pubes, I feel
a distinct wave communicated to my other index finger in
the bowel, I know that to be a point at which the trocar
can be inserted with safety. Also, if I find a well-marked
rounded tumour over the pubes which is dull on percussion,
the bowel around it clear and distinct, I have reason to
believe that an operation over the pubes would be successful.
Again, the question arises : why not attempt to relieve the
bladder by an operation on the urethra itself from the
perineum, so as, if possible, to cure the stricture, and at the
same time relieve the bladder ? Might it not be wise in this
manner, as it were, to kill two birds with one stone, and not
be content with merely puncturing the bladder by the rectum
or above the pubes ?

At this point let me revert to the different practice and different experience of surgeons in relation to this matter. Let me give you the experience of Mr. Liston. He once said, from his chair, that during the whole of his connection with the Royal Infirmary of Edinburgh, and subsequently with this hospital up to the time at which he spake — namely, three or four years before his death—he had never punctured a bladder for retention of urine. On the other hand, there are men living in this town who have punctured a bladder fifty times or more. Mr. Liston meant to imply that a good surgeon ought rarely to find it necessary to resort to any other means than the catheter in circumstances of retention. But do not suppose for a moment that the gentleman I spoke of who has punctured a bladder fifty times, does so because he fails to pass the catheter under those circumstances, but because he thinks it wiser to puncture the bladder than to persevere too much with the catheter. Then, again, both Liston and Guthrie occasionally performed the perineal operation just spoken of. From the perineum the urethra *may* be reached behind the stricture. Now, without entering into a long discussion on the subject, I may say that this mode has lost favour of late years. It is no easy thing to find the urethra behind the stricture; and a man may make an awkward wound in the perineum, and never hit the urethra at all. Then it does not follow that it should be desirable to divide the obstruction at all, so far as its cure is concerned; for the stricture, when the time comes to treat it, may be amenable to dilatation. The reason why puncturing through the rectum has been done so often by Mr. Cock, of Guy's Hospital, is because he conceived it to be an excellent kind of treatment. He says: "Let us withdraw the urine from the urethra

G 2

altogether for a few days, and the urethra will recover itself, so that we may be able to cure the stricture with ease." And that is often true. He punctures the bladder by the rectum under the circumstances I have mentioned ; and this is his instrument for doing it. In this way, the water not passing by the urethra at all, the urethra is lying fallow, so to speak ; and in a short time the instrument can be passed —say No. 2, or 3, or 4 catheter, although before you could not pass No. 1. This, then, is a species of treatment of stricture which Mr. Cock has introduced ; and, at all events, he has proved how easily and safely this operation may be performed: he has, in fact, familiarized us with a proceeding which before was often thought a very grave and serious affair.

If, then, you have failed to pass the catheter, and the symptoms are urgent, you have two proceedings before you : puncturing by the rectum, or above the pubes (*see* Fig. 14). By carrying into the bowel your finger—a reasonably long finger—you arrive at a point just behind the prostate. The other hand is placed above the pubes, that the wave of fluid, by its pressure, may be distinctly felt by the finger in the rectum. You are then quite certain of what you are going to do. Along this finger, kept firmly in place, your trocar is slipped, and then boldly, but carefully, pushed into the bladder. This is always an anxious moment, because, if you have not hit the bladder, it is a serious matter to have thrust this long instrument into the centre of the body, and find no urine escaping. The best position for the patient is sitting on the edge of a bed, resting his back on pillows behind, the legs apart, each on a chair; an assistant by him placing one hand on either side above the pubes, so as to steady the bladder, and press it down towards the rectum.

It is well to remember, subsequently, that if the canula slips out, you will not be able to get it into the same opening again. The muscular fibre of the bladder instantly closes, and you have to make another puncture—not a matter of much consequence, but better avoided.

FIG. 14.—A section of the pelvis showing bladder and rectum.

For the supra-pubic operation you divide the structures in the middle line until you reach the linea alba. Then carefully making your way deeper, you will soon discover fluctuation ; and, having the bladder steadied as before, you will thrust the trocar slightly downwards. In this case you retain the silver canula for two or three days ; but you may soon substitute a gum instrument. Now, supposing there is a probability that your patient may require this artificial relief by tube for some time, you will, of course, prefer the situation above the pubes, because it is much more easy to wear the tube there than it is in the rectum, where it interferes with the functions of the bowel, and is otherwise much in the way. I have known patients who have passed all the urine through a tube above the pubes from ten to fifteen years, and who lead active and comfortable lives in consequence, the natural passage being completely obstructed. One of them, who had suffered greatly before, and was now in perfect comfort, told me that "he did not know whether this mode of passing water was not preferable to the original one !" That, however, I conceive, is quite a matter of taste.

Our next subject will be Extravasation and Fistulæ.

LECTURE VI.

EXTRAVASATION OF URINE AND URINARY FISTULÆ.

BEFORE commencing the subject of urinary fistulæ, I shall briefly allude to a condition closely related to retention, just considered—viz. extravasation of urine. And it is related in this way:—Suppose that from bad treatment, neglect, or otherwise, the patient has had no relief, and when you are called in you find there is no question of puncturing the bladder, for the urethra has punctured itself, so to speak ; that is, Nature (as in the case of all maladies) has done something ; often she acts in a clumsy way, but sometimes not more clumsily than the surgeon. There is no question that many who are the subjects of stricture or retention, and receive no surgical aid at all, will lose their lives in consequence ; but they are sometimes saved by extravasation of urine taking place. What then happens is, that behind the stricture a portion of the urethra gives way, perhaps during some violent act of straining, and through the rent so made, a quantity of urine is driven with great force into the cellular interspaces. Well, we know where that must go, from the anatomical disposition of the fascia—viz. into the scrotum, up into the groin above Poupart's ligament, and towards the belly. Occurring, as it usually does, in the bulbous part of the urethra, where the walls of the canal are weak, the urine cannot pass backwards behind the scrotum, or the back part of the perineum ; neither can it get into the thighs, because it is checked by Poupart's ligament. I have seen it rise as high as the chest, and I have made incisions to evacuate it, in a bad case of extravasation, as high as this point. Once taken place, it follows that at every action of the bladder more fluid is driven in with force, so that the cellular interspaces are separated, and the fluid usually finds its way upwards. Generally, you ought to know at once, by the appearance of the patient, what has taken place, although it is possible sometimes to confound the condition I speak of with ordinary inflammatory œdema of the scrotum, for it may commence very gradually and insidiously. In ordinary cases, you see a hard perineum, a large red tense scrotum, the penis swollen, and a red blush perhaps rising over the pubes. In order to ascertain the true state of things, you must ask for the antecedents, and you will probably learn that there was great difficulty in passing urine, followed by rather sudden relief. When a man has had retention for some days, and extravasation suddenly takes place, instant relief is experienced—the frightful want to make water disappears as soon as the fluid finds its way into the scrotum; but he soon feels new pains, not necessarily very severe, and, what is worse, constitutional symptoms rapidly set in. The poisonous fluid quickly destroys the cellular tissue, so that sloughing soon begins. After forty-eight hours or so, gangrenous discolorations appear, and the urine may find its way into the corpus cavernosum, when a dark spot appears on the glans, showing that the structure of the penis itself is infiltrated. Without describing further this condition, which you must have seen for yourselves, and which may be seen now to some extent in a patient in the ward upstairs, let me say, do not in such a case be afraid of the knife. You have no occasion for a catheter; the urine has found its way into the cellular tissue, and you must let it out as freely as possible. On each side of the perineum make a deep inci sion. You need not limit the incision to two or even three inches; because you are really cutting into urine, not into flesh. The structure is so enormously distended, that there is but little flesh to divide; and although the incision may appear very deep and long, when the water has run out, it will be comparatively small. The incisions generally bleed rather freely. One may soon lose a pint of blood from three or four incisions. The urine runs out also, and as the distension goes off, the vessels are enabled to contract better; but if you see any little vessel spouting, tie it at once. An incision should be made on each side of the penis, because if it is made in the middle line, there is not sufficient communication for the incisions on the one side to relieve the other. Do not be extravagant in these incisions; still it is better to err on the side of freedom than to be too niggardly in the use of the knife. The next day, if the case has done well, you will find the scrotum much reduced in size, and the parts altogether much less swollen and inflamed. You have now a direct communication through the cellular tissue from the bladder, and, with that rent behind the stricture, as a rule, you will be safer in letting the catheter alone, and permitting the water to drain off. What happens? Why, just what happens after puncturing the bladder. When the water flows off by another passage, the urethra begins to improve, and in three or four days you will probably have no difficulty in passing No. 3 or 4 catheter. With these patients, bad as they are, prostrated as they are when you see them, if the case has not gone too far, and too much gangrene has not taken place, very striking and rapid recoveries often follow. The whole scrotum may slough away, and the testicles may be seen uncovered in the wound, and yet all may heal up soundly and well.

This leads us to the next subject. After the exit of the urine by these artificial channels, some of them fail to heal and remain patent, and thus form what we term urinary fistulæ. Now this day week there were in my ward three cases illustrating this condition, and rather exceptionally obstinate ones. One was caused by extravasation of urine, and the other two by the more usual and common cause— stricture of the urethra.

We have spoken of extravasation ; how does stricture give rise to a fistula ? Thus: in some patients, when a stricture has existed some time, and has had no treatment, or bad treatment, it is not uncommon for chronic abscess to form, say in the perineum, between the urethra and the surface. In time this opens externally, and a few days afterwards a little urine finds its way through it, and passes at each time of making water. If no relief is afforded to the patient, another forms ; and so several sinuses may arise, and other openings in various and surrounding parts, all giving exit to the urine. These fistulæ may take place in a great number of situations, such as in the penis, in the scrotum, in the perineum, in the groin, and in the rectum ; but the last are very rare. We will consider the first four by themselves, because fistulæ which open into the rectum require a different treatment altogether. Then, respecting their specific charac- ters, I shall make three classes, examples of which may be found in any of the localities named. First, fistulæ may consist of simple openings or channels between the urethra and the external surface; or, secondly, they may be sur- rounded with inflammatory induration, which makes them more difficult to heal ; and, lastly, they may be complicated with loss of substance from sloughing, so that even a portion of the urethaa itself may have been destroyed ; and these are the most difficult cases to deal with. Hence these fistulæ classify themselves naturally as simple, indurated, and fistulæ with loss of substance.

I will deal briefly with the " simple urinary fistula." Whatever part of the canal it is connected with, it almost always heals, if the stricture with which it is associated is dilated. Dilate the stricture, and in nine cases out of ten the fistula will heal. Patients, especially in private practice, are often extremely anxious about the result of an opening in the perineum or elsewhere through which the urine passes ; and it is right that you should assure them for the most part, if the stricture is thoroughly dilated, the unnatural passage will heal of itself. But there is another point to be considered in connection with this—viz. the quantity of urine passed by the fistula, compared with the quantity passed by the natural passage. Of course the gravity of the thing depends very much on the relative proportions passed by the two ways. Usually about three-fourths of the water will pass by the right way, and one-fourth or one-fifth by the wrong passage. If a large quantity—say three-fourths —passes by the unnatural opening, then probably a considerable amount of stricture is present. Nevertheless, as the stricture is dilated, you will see that the proportion of urine passing through the fistula will gradually decrease until it ceases altogether, and the part soundly heals ; but this latter result is achieved only on the condition that you maintain the stricture in a dilated condition.

Now we come to fistulæ which are associated with much inflammation and induration in the perineum. These may be multiple ; in which case you may find five or six openings. I have known a man with a dozen, so that instead of the urine being discharged by one stream, it flows as from a watering-pot. Even this condition, however, very much improves as the stricture is dilated, and may get quite well; but this is not always the case. Then there are some instances looking much less formidable, like those upstairs, with perhaps only two or three openings, through which the greater part of the water has passed for a long while. You recollect that we fully dilated the stricture in each case; but still no improvement as regards the flow of urine through the fistulæ took place. A No. 12 catheter was passed; but the patients did not get well. The condition of the perineum improved very much; but still more than one-half the urine obstinately held its erratic course through the perineal open-ings. Now, what is commonly done in such circumstances? Usually operative proceedings of some kind are resorted to; or, if these have been postponed or rejected, a rather tedious process has been employed. The principle laid down—and I have myself applied it successfully—is, that it is necessary to take care that the external openings should be very free, ensuring this either by means of the knife, or by potassa fusa, or by some other means; so that the urine may not be detained in its way from the urethra to the external surface, causing fresh induration or thickening. Next, you may go on to excite adhesive inflammation in the track of the fistula by a hot wire, or by touching it with cantharides or a strong solution of nitrate of silver. No doubt this treatment sometimes succeeds; but it is at best a tedious process. Then it was sometimes attempted to cure such fistulæ by tying in a gum catheter for weeks, or even for months. But this generally fails; and for this reason: that urine always finds its way from the bladder by the side of the catheter, along the urethra, and so into the fistulæ by the force of capillary attraction; and thus the object supposed to be attainable never was and never could be so accomplished. The practical surgeon soon discovers that tying in an instrument never ensures the transit of the urine through it; some will always pass by the side and defeat your purpose. I have, therefore, adopted the plan of teaching the patient to pass the catheter himself; and that is by far the most rapid and the most certain method. With regard to the two cases upstairs, ten or fifteen years ago I should have applied potassa fusa or the galvanic cautery, or something of that kind; but the fistulæ have soundly healed through ensuring,.by means of the catheter, not, as by the other process, that the urine should percolate rapidly through the perineum, but that it should not pass through at all—in fact, turning the current another way. You first teach the man to pass a No. 7 or 8 gum catheter himself—an easy matter enough. He then agrees to pass it every time he requires micturition, night and day. On no occasion is he to permit the urine to flow spontaneously—say during five or six weeks—not even when he goes to stool; and this is avoided by always using the catheter immediately before. That plan has been followed in each one of the three cases in question, without difficulty and with perfect success; for each man has a sound perineum, and has now relinquished the use of the instrument.

Now I come to the third form of fistula, that in which there is loss of substance. This class must be dismissed rather briefly, because its full consideration would involve a tedious detail of many different surgical procedures. Where you have this loss of substance, a plastic operation of some kind is generally required, to fill up the gap which exists.

When the opening is small, you may contract it very materially by the heated wire or galvanic cautery, or by any mode
which tends to produce a contraction of the tissues. You know
that cicatrices which result from burns contract considerably,
and you avail yourself of that action in this instance. Most
commonly, however, if the soft parts have been largely
destroyed, some plastic operation is required for the cure.
Thus, on passing a silver catheter, when a portion of the
urethra has sloughed away, you may see perhaps a quarter,
or a third, or even half of an inch of the catheter exposed
in the wound. The successful treatment of such cases
demands much care and nice management. They do not
often come under our notice, and less often do they get
completely cured. I have had in the hospital but three or
four such cases, in which, by means of plastic operations,
the patients have been entirely cured. Some of you saw
one last winter—a man who had just between the angle of
the penis and the scrotum an opening, showing at least a
third of an inch of the catheter, the whole of the floor of
the urethra having sloughed away. The operation in that
case was one of the most successful I ever saw. The first
operation completed it, with the exception of an opening
not larger than a pin-hole. What was done was to pare the
edges all round, then to get a flap of skin from the scrotum
below, which was brought up to cover in completely the
wound, the margins being carefully attached by a number
of little sutures. That fistula healed perfectly. And why
did it heal? Here is the important point: there was one
condition necessary, without which it would have failed. A
week or two before this operation, I made the patient learn to
pass the catheter habitually, so as to draw off every drop of
urine; and finding him thoroughly expert at it, I performed

the operation ; and for a month he never allowed a single drop of water to pass otherwise than by the catheter. Had I tied the catheter in, it would not have been sufficient, because the water, as I have told you, always finds its way by the side sooner or later. Luckily, he performed his part of the compact to the letter for the stated term, so there was no reason why the wound should not heal there as well as anywhere else. The little tiny opening which remained was perfectly closed with the heated wire, and the urethra can now perform all its functions perfectly well. You know there is another very important function connected with this canal, besides that of micturition. I do not know what you may think that function worth, but it is one which may involve very important considerations in cases where the transmission of a great family name or title or estates depends on it. Whether this be so or no, there is no doubt that every man conceives that to be an important function for himself, whatever others think of it for him ; and it could not in this case have been performed unless that opening had been closed.

Now, in order to go into the whole subject, I should want a lecture or two to tell you of the different kinds of operations which are performed in different spots. I have taken the case described as a typical one. It is one of the most difficult to close. The penis is liable to differences in form : the patient may be troubled with erections, which may damage any operation, and there is very little flesh to deal with. In the perineum you have two or three inches of depth, so that you can cut flaps of any size and thickness.

A word or two about the rectum. You remember that I made an exception in reference to fistula coming from the urethra into the rectum. There is a case upstairs in which it occurred from the patient himself thrusting a bougie from his urethra into the rectum. More commonly it occurs from prostatic abscess. In these cases, at each act of micturition, urine passes into the rectum—often a very troublesome, even distressing circumstance. The rectum becomes excoriated, and the patient is obliged very frequently to go to stool. I shall say very few words about it, because each case must be treated on its own merits. I will give you the result of my own experience, and that is all I can do. I do not know that there are any published records respecting these cases. They are very few in number, but they are very important when they do occur. I cured one case by position. It was the case of a young officer whom I saw in private practice (I have not met with a precisely similar one in the hospital, and therefore refer to it), who passed three or four tablespoonfuls of urine into the bowels at each act of micturition, after having had some abscesses there which I did not see. It occurred to me, after some wholly inadequate treatment by other means, to tell him to lie down on his face and make water in that position, never allowing a drop of urine to pass in any other way. In a few weeks he was quite cured— very luckily for him, and for me too. If you ever meet with such a case, the plan is worth trying. I have had two other such cases since, but the plan has not succeeded. In the instance to which I refer, it occurred to me that the force of gravity would carry all the proper way, and it did so. None passed into the rectum, and at the end of six weeks the patient was well. I saw him some years afterwards, and he was soundly cured. I take it that I should now have made that man pass the catheter into his bladder and draw off every drop of water: and I have no doubt it would have been successful. Unless there is a loss of substance, that usually does cure the patient; but if there is a loss of substance, and worse still, if the opening is from the bladder into the rectum direct, then nothing is left but to examine the place thoroughly in the first instance. Put the patient on his back, as for lithotomy, and introduce a duck-bill vaginal speculum, as you have seen me do, so as to get a good light thrown upon it. If it is sufficiently large to do a plastic operation, I should not hesitate to perform the same operation that is done for openings between the bladder and vagina—that is, to pare the edges, and stitch them together with silver sutures,—only it is more difficult, as there is less room in the rectum for manipulating than in the vagina. In the vagina there is plenty of space for the work, but I have done it also in the male in one case. I found it possible to do, though difficult; and I believe that is the best plan when these cases occur with loss of substance. If the opening, however, is very small, it may be greatly diminished in size, if not closed, by applications of the galvanic cautery.

II

LECTURE VII.

STONE IN THE BLADDER.

GENTLEMEN,—I wish to give you to-day a sketch in outline, embracing all the principal points, if I can, of a very large and important subject—viz. stone in the bladder of the adult male. I shall say little about stone in children, and nothing, at the present time, about stone in women.

First of all, in what classes of cases is stone most common ? Contrary to what is stated in books on the subject, it is most common in individuals from fifty to seventy. In books you will find it stated to be most common in children. Perhaps it is so if you take the number of cases in children as compared with the number of cases in elderly adults, although this is by no means certain, but not if you take the individuals of either class relatively to the numbers of that class living at the time.

I think it may be said that the most favourite period for calculus is from about fifty-five to seventy-five ; the next in order is that below puberty, and the most rare period is that of middle age. Looking at the cases numerically, you may put it down as a general rule that half the total number of hospital cases occur below thirteen years of age. I cannot refer you to any more exact researches than those which were made with great labour some years ago by myself. Out of 1,827 such cases, each one of which was reported to me in writing, and of which I knew all the principal particulars, one-half occurred before the age of thirteen. You will remember that this is in hospital practice, which gives a somewhat different set of cases from those seen in private.

Now, with regard to the varieties of stone, I do not give you here all the chemical distinctions; that is not necessary; but I will mention three chief classes which it is important to consider in relation to the practical management and the removal of stone. That which is most frequently met with is uric acid and its combinations; the second is that in which phosphoric acid is combined with volatile alkali and the alkaline earths; and, lastly, there is oxalate of lime. For all practical purposes those are the three great divisions. Among these, uric acid and the urates form about three-fifths in number, the rest being phosphates, with the exception of about three or four per cent. of oxalate-of-lime calculus.

Next, what is the ordinary history of stone? You know well, of course, that the appearance of a stone of some size in the bladder is not the first stage of the malady. The stone begins always—I am now speaking of uric acid—as fine sand or gravel, to use popular terms: that is to say, there is an excess of urates, which, perhaps, steadily persist; then, possibly, of uric acid, in characteristic cayenne-pepper-like masses of crystals; then there are small rounded bodies of this latter aggregated within the kidney, about the size of shot, or somewhat larger, of which you have very good specimens here. An acid calculus, then, is always formed within the kidney; and it is a fortunate thing if it descends into the bladder, because sometimes it is retained in the kidney, and, as renal calculus, becomes the source of great misery, for which surgery can do little, although medicine may be of some small service. But if it comes down into the bladder, it is usually passed—say in nine cases out

H 2

of ten—without any operation whatever. The patient has
an attack of severe pain in the back, over the hip, in the
groin and testicle, lasting for some hours, and it probably
ends in the descent of the calculus from the kidney to the
bladder. When it gets there, after a day or two, or before,
it is generally expelled by the urine, and there is an end of
the matter; only the patient must be told, or ought to know,
that this occurrence shows a strong proclivity towards the
formation of stone, and he should immediately do all he can
to prevent its continuing, as in the nature of things it will
do. But if the bladder is unable to expel the calculus, it
soon increases in size by deposit on its surface of acid from
the urine, and a very hard but rather brittle stone is formed
in the course of time. All the stones which you see in this
box have been passed through the urethra by the natural
efforts; and it is worth knowing how large a stone may be
so passed in some cases. Usually, when they get as large
as some of these, they fail to pass, and then some operation
must be performed for their removal.

The phosphatic calculus is not necessarily formed in the
kidney; it is so sometimes, but usually it is formed in the
bladder. In the mucus of a diseased bladder a good deal
of phosphate of lime is formed, and this, meeting with ammonia from the decomposed urine, produces the ammoniaco-
magnesian phosphate. This, with the phosphate of lime,
makes "fusible calculus," the commonest form met with.
Its structure is not dense, and it is easily crushed. The
oxalate of lime, or mulberry calculus, I need not tell you, is
not originally formed in the bladder, but in the kidney, and
it is the hardest in structure and the roughest in external
surface of all.

Now, what are the symptoms of stone? These we will

seek, if you please, by means of the four questions always to be used. A patient tells you, perhaps, that he has passed some gravel for a year or two, and may show you some small stones which he has passed. For the last few months, perhaps he has not seen any; but, during that period, has had increasing difficulty in passing his water, a circumstance which will strongly excite your suspicion. First, as to frequency of micturition. The patient has for some time had more or less frequency, but it is increased during the day when moving about, and decreased at night, when he is at rest. This is contrary to what usually takes place in prostatic enlargement, and hence it is a good diagnostic point.

You next ask for pain. The patient with calculus of the bladder has almost always pain, and this is referred to a particular spot, the lower part of the glans penis, about an inch or less from the external meatus. Remember, you may have that pain when there is no stone in the bladder, as in chronic prostatitis and in some affections of the bladder; but in stone it is almost always present. Further, with regard to pain, the question should be asked, whether the patient feels it before, during, or after making water. He will tell you that it is during and after; whereas you know, in hypertrophied prostate, and in all cases in which water is retained, the pain is before passing water, and it is relieved by the act. A man with stone feels pain after making water, because the foreign body is then left in contact with the lining of the bladder, and being driven to its neck, severe pain and a sense of desire to micturate are felt, perhaps for four or five minutes, until fresh urine entering, the coats of the bladder are separated from the stone.

Then you ask him the condition of the urine. In nine cases out of ten you find that there is muco-pus, and also, perhaps, streaks of blood. There is almost always more or less clouded or muco-purulent urine with calculus.

Then, lastly, you ask him with regard to blood. Almost invariably at some time blood has passed, and it is increased by movement. A patient with calculus suffers more pain, and has more bleeding, and the urine becomes more thick, if he takes much exercise. He cannot ride on horseback, nor in a jolting vehicle, without considerable pain ; in fact, all quick movements of the body intensify the symptoms. Certainly a patient presenting such conditions ought not to be allowed to go without sounding.

Then, how do you sound? You should employ an instrument like this, with a small short curved beak, because it can be turned in any direction. If you take an instrument with a large curve, like a catheter, you are unable to rotate it in the bladder, and hence it does not explore sufficiently.

When I entered this room, you heard me ask for the hospital sounds, for I knew I should find among them a good example of what a sound ought *not* to be. Here is one, for example, which no one could rotate, or ever find a small stone behind an enlarged prostate with, except by sheer accident. It has precisely the form of a common catheter. You will say, naturally enough, "Why are such sounds here, and who has used them ?" They were used formerly, and found a good many stones, too, in the hands of our illustrious predecessors. But I will answer for it, they have missed a good many stones, also ; and this is precisely what I want you not to do. I have no hesitation whatever in saying that more stones are missed in sounding than are found by the ordinary methods adopted in this country ; and that must be the case if a sound of the form of the

common catheter is relied on for the purpose. But with an
instrument which has this small beak at the end of it, you
can search in every direction (*see* Fig. 15). If there is a large
stone, of course you can find it with anything ; but our

FIG. 15.

great object is to find the small stones. As a rule, anybody
can find a big stone ; the art consists in finding a small one·
It is most important to find a small stone, because it will
grow large, and may be very formidable to deal with ;
whereas, when it is small, it is a far less formidable matter.
You may promise the patient, in the case of a small stone,
that it may be removed without risking his life ; whereas in
the case of a large stone, there is always some risk, often
considerable danger.

In the next place, how are you to use this instrument ?

First of all, it is not to be introduced in the same way as
the ordinary catheter. With the ordinary catheter you stand
at the left side of the patient, and make a gentle sweep
thus into the bladder. With the sound you stand at the
right side, and use a different manipulation, which I shall
show you on the living patient, at my next lecture, post-
poning, therefore, my remarks on the subject until then.

But you have something else to do besides merely discovering the presence of stone. It is necessary to have other

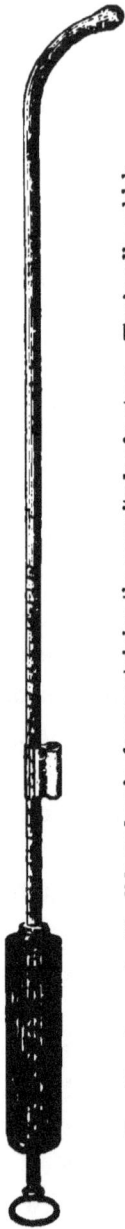

FIG. 10.—A sound with slide and scale, for ascertaining the magnitude of a stone, but smaller, affords great facility in sounding. The handle, which resembles that of my lithotrite, but smaller, affords great facility in sounding.

particulars respecting it, because the nature of the operation to be performed will depend on them. First of all, it is essential to know what the size of the stone is before you decide on what you will do with it. From the note elicited by merely striking it, you can get some indication of its size. It is often sufficient for practical purposes to use a sound (which I have long used myself, and have recently introduced) provided with a little slide on the shaft, which, by proper manipulation, enables you to ascertain very nearly the size of a stone, as you have frequently seen in the wards.

Then there is another way. You may introduce a lithotrite (which gives, however, a little more disturbance to the patient) and seize the stone in two or three directions, so as to ascertain its diameters.

At the same time you ascertain its nature. A phosphatic stone gives a very different sound from the others. The specimen before me is dry, and therefore will not give the sound to which I refer. When wet, it is spongy and soft, with a rough surface, and always gives a dull note when struck; whereas the uric-acid stone gives a hard ring. Then you will judge partly by the condition of the urine. If the urine is acid, and if, also, uric acid is thrown down, you may conclude that the patient has a uric-acid stone. If so, it is likely he has passed small calculi before. If the urine is very alkaline, and there is a great deal of phosphatic matter, you may conclude that it is a phosphatic stone, or, at all events, it is covered with phosphates.

The number of stones is the next thing. Usually there is only one stone, but occasionally there are more. There is a patient here on whom I shall perform lithotrity to-morrow, who has two rather large uric-acid stones in the bladder. The way to determine that point is this: Having seized one in the lithotrite, you move it gently in every direction as a sound for others. If then you encounter one on one side and one on the other, you know that there must be at least three stones.

I have spoken of uric-acid and of phosphatic calculi. But it may happen that you have an oxalate-of-lime stone —a very important thing to ascertain. You examine the urine, and see if there is much oxalate of lime thrown down. The patient may have passed a small mass of oxalate of lime before, and you may infer that an oxalate-of-lime calculus exists now; but then it may be covered up with phosphates, and so deceive you. I will give you a case in point. I had to deal with a very large stone in the bladder of a private patient. I crushed the stone four times, bringing away large quantities of phosphatic material. I always noticed that my lithotrite never went through the stone; it always went a certain way, and then there was a hard mass. After four sittings I could not crush any more. It was clear that there was a very hard central stone, from which I had taken away a large crust of phosphates. My strongest lithotrite made no impression upon the stone. I know, from experience, the recoil of the lithotrite from an oxalate-of-lime stone so well, that I had no hesitation in saying this was an example of that kind. Accordingly I cut him, and a large oxalate-of-lime stone was removed. In a case like that, you would not have oxalate of lime in the urine, but phosphates. In the case of a uric-acid stone, you may make an impression upon it with the lithotrite—its teeth will bite into it; but in the case of a large oxalate-of-lime stone, it is like laying hold of a piece of iron—you make no impression whatever upon it.

Having got all these data, the next important question is, what are you to do ? Are you to cut or to crush? You know there are only two modes of removing the stone. You must either make an opening sufficiently large to admit of its withdrawal, or you must crush the stone into very small fragments, so that they may be expelled by the natural passage. It was less important to make a diagnosis of all these points, when we had but one operation—namely, that of cutting. Formerly, whether the stone was large or small, the patient was always cut. There was no other way of removing it. Now that we have two operations, it is very important that we should choose the right one ; because, let me tell you, if you do not determine pretty accurately the characters of the stone, and select the right operation, you may do more harm than if you cut every patient. If you crush the very large stone, and cut for the very small one, you will have greater mortality than if you simply resorted to the one operation of cutting in all cases. When lithotrity was first introduced, it was rather a clumsy operation ; and when the cases were not judiciously selected, when surgeons crushed without making a diagnosis of all these points—crushed stones that ought really to have been cut, and left for cutting stones which might have been crushed

—the entire mortality resulting from operations for stone was greater than previously, when every case was cut. I cannot give you a stronger argument for the necessity of apportioning the operations judiciously.

Now, without taking up your time too much, I will lay down what you will understand to be the axioms which should direct you, in a general way, in making your selection. First of all, I will say that all stones, under puberty, with very few exceptions, are to be cut. Under fourteen or fifteen years of age, stones occurring in the male are to be cut, unless they are very small, and can be crushed, say, in one operation; because lithotrity is not a very easy or successful operation in children, the urethra being small and the bladder very irritable; whereas, as is well known, lithotomy is a very successful operation in these cases. We do not want a better operation, comparatively speaking, and one may be content to let well alone. Not more than one death in fifteen or sixteen cases occurs from lithotomy in children. I do not think, therefore, we can do better than to cut in these cases, as a rule. If, however, you have in a child of, say three or four years old or upwards, a stone no bigger than an orange-pip, you may very probably succeed in crushing it, under chloroform, in one, or at most two sittings; and this it is usually advisable to do.

That leaves us all the cases above puberty. Then I will say, in general terms, that all the cases above puberty are to be crushed, with certain rare exceptions. The first exception is in a case of an oxalate-of-lime calculus, which is, let us say, an inch in diameter. Under an inch in diameter you may crush an oxalate-of-lime calculus. I have crushed four or five in my time. The cases are very rare. Two of them were in this hospital. An oxalate-of-lime stone, from the size of a bean up to an inch in diameter, can usually be crushed; above that size no instrument can deal with it, and the fragments will be so hard that the operation might be of doubtful value, even if we succeeded in crushing. That, then, is the first exception to the general law that all cases in adults are to be crushed.

Secondly, a large stone of uric acid, or a phosphatic stone of very large size, had generally better be cut than crushed. Mechanically speaking, it is possible to crush any stone, whether uric acid or phosphatic; but, considering the number of sittings required, and the amount of irritation produced, the single operation of cutting will be the better of the two when the stone is—shall I say—full two inches in diameter. A stone which is two inches in diameter, either phosphatic or uric acid, had perhaps better be cut. No doubt a rather larger phosphatic stone may be crushed. Here is a very large one, two ounces and a half in weight. The phosphatic stone is very friable, and you may deal with rather larger phosphatic stones than uric-acid stones by lithotrity. So much for the exceptions to cutting, regarding the characters of the *stone* itself.

Now what are the conditions on the part of the *organs* which will make it necessary for you to cut instead of crush? If you have a bad stricture of the urethra you cannot crush; if you have a very highly diseased condition of the bladder, probably you cannot crush; but such exceptions are very few indeed. First of all, I will tell you what are not exceptions, but which are stated to be so in books, and are generally considered as exceptions; because lithotrity has advanced since the time of most books. I have recently crushed a uric-acid stone in a case of organic stricture with small instruments made for the purpose; but the stricture was not a very narrow one. Hypertrophy of the prostate was said to be an exception : it was said that you could not crush under those circumstances. I make no difference whatever in respect of that matter ; I would as soon crush in the case of hypertrophied prostate as in any other. It is only a question of delicate manipulation. If the hypertrophied prostate occurs in a man who has had instruments passed, he will have become habituated to them, and is therefore a better subject than a healthy one who has not been so accustomed. Next, it was said that when the bladder could not empty its contents by its own power, and the urine had to be drawn off by means of a catheter, lithotrity was contraindicated, inasmuch as the fragments could not be passed. On the contrary, I rather prefer such a case, for the reason just assigned—that is, the bladder and the urethra are habituated to instruments ; and as to not removing the fragments, there is no difficulty at all in removing the very last fragment, thanks to the improved methods now employed. It was said also that great irritability of the bladder was a reason why we could not crush. It was said that if the bladder could not hold above three or four or five ounces of urine, there would be no room for the lithotrite to work, and therefore the surgeon must cut. I make no objection on that ground, because the irritation of the bladder is due to the presence of stone; and as soon as you begin to get away the stone, the irritation often diminishes. Besides, it is not necessary to have four ounces of water in the bladder: one ounce is ample. There is no occasion to have four or five ounces in order to perform the operation of lithotrity. It might have been so with the old clumsy instrument, but with modern instruments there is no necessity to protect the bladder from contact with them. When instruments were used that were apt to catch the coats of the bladder, it was no doubt desirable to have a quantity of water in the bladder; but with modern instruments, which will not lay hold of the coats of the bladder, there is no difficulty whatever in crushing with a single ounce of water. I do not care whether the bladder is empty or contains a large quantity of water, provided only that it is not too full. Nothing is worse than too much water, because the stone rolls about, and you must, figuratively speaking, play a game of hide-and-seek to catch it. It is better to have an empty bladder than a bladder with half a pint of water in it.

You see, then, that the exceptions are very few; indeed there are very few adult cases which cannot be crushed, provided you give proper care and attention. If surgeons of the present generation now growing progress, as they must, and become more intelligent and more careful than those who have gone before—if they are better acquainted with the subject, as in the nature of things they must be, as our sons will be wiser than ourselves, and our grandsons wiser than they, there will be fewer and fewer exceptions; because, if the stone is discovered when sufficiently small, *it can always be crushed with an almost* CERTAIN CHANCE *of success;* so that lithotomy for adults must at some day disappear, except for cases which have been neglected by the patients themselves, or have been overlooked by the medical attendant.

A rather large uric-acid stone is the growth of several years; a large phosphatic stone is perhaps the growth of two or three years; an oxalate-of-lime of full size, say from seven to ten years; and it is very hard if, long before the expiration of such periods, the stone cannot be found and disposed of by lithotrity. It is certain that if there be proper intelligence and proper supervision of the patient, a stone would be always discovered, when it can be crushed with almost a certainty of success; so that the only cases in which lithotomy will have to be performed will be those in which the patient has neglected himself, and, although suffering from severe pain for years, has never gone to any surgeon to tell him about it. But those cases must be very few indeed. I hope you will live to see the day when lithotomy for adults will disappear. I do not suppose I shall; but I do expect to live to see one thing, and that is, lithotomy becoming very much rarer than it now is. You certainly will live to see it one of the rarest operations. I do not say that I look forward to that with any particular pleasure; for it is a grand operation, demanding all the skill, self-command, and force of a man. It is one of the best practical tests of a good surgeon, and, looking at it from that point of view, one cannot desire its discontinuance; but it will disappear, most assuredly; and as it will be for the benefit of humanity that it should, we must acquiesce in the result.

Next week I shall place two stone patients on this table, and demonstrate the operation of lithotrity for you here in my lecture on that subject.

LECTURE VIII.

LITHOTRITY.

GENTLEMEN,—I shall place on our lecture-table two patients
with stone in the bladder, one of sixty-two, the other of
sixty-five years of age. One man has a stone about an
inch in diameter; the other has two stones, each about
three-quarters of an inch in diameter; and in both the
composition is uric acid.

Now, if a patient has never had any instrument passed
into the bladder before, and if the urethra is not capacious,
it is desirable to pass a bougie two or three times before
commencing. In the present cases the urethra is not very
sensitive; it is large in size, and there is no occasion to do
this. Next, it is desirable that the patient should not be
below his ordinary standard of health. You will not begin
to operate when there is an attack of fever, nor unless the
digestion is in fair order, and the bowels are acting tolerably
well. Take care that you have the local organs, and the
whole system, in as favourable a state as possible at the
outset.

Having decided to crush, there arises the question of
instruments. Now, I shall simply show you the kind of
instrument which I have used, and also, to a certain extent,
what has been used in the past. I may say that lithotrity,
as an operation, owes its existence to the French surgeons,
mainly to Civiale, but not forgetting Leroy d'Etiolles and

others. My old friend Civiale, who only last year died at a good old age, and full of honours, was the first surgeon to crush a stone successfully; that was in the year 1822. It had been done occasionally by patients themselves. In one case a man had managed to grind down with a small file a little stone in the bladder, and that has been called lithotrity. But the first man who systematically performed the operation on the living patient was Civiale, and he operated on his first two patients before the Academy of Medicine, with this instrument that I hold in my hand. You see how different it is from anything we now employ. It is a straight instrument, with a central axis and three claws, which were made to project after its introduction into the bladder. [The manner of using it is shown.] You see what a very different mode of proceeding that is from the method now adopted. Yet that was, to a certain extent, a successful operation. I cannot now carry you through all the stages by which the various improvements originated, but the next great change was the curved instrument. This was a great step in advance, and the curved form, with slight modification, is still the prevailing form with all operators. But here is an old pattern, worked with a simple thumb-screw in the handle, which still holds its ground, I regret to say, in this country. It has long been exploded elsewhere, although still used in London. Nevertheless, it is the instrument with which Sir B. Brodie earned his success, and is used thus. [Explanation.] You see how long a time it takes, and how much movement is necessary with it in the bladder. You cannot do any respectable amount of crushing in less than five minutes in this way. A great improvement upon that instrument was made by Civiale and Charrière, of Paris. The object was to avoid the loss of time and the jar occasioned

I

by screwing and unscrewing, and in this new instrument the sliding movement is instantaneously changed into a screw movement, and *vice versâ*, by the turning of a disc in the handle. [Instrument and method exhibited.] Meantime Sir William Fergusson devised and adopted the rack-and-pinion method, which is an improvement on the old instruments before described. Lastly, I show you a lithotrite, due in part to my own design, and in part to Messrs. Weiss, which is now so widely used that I must notice it also, especially as it is that which you always see me use here.

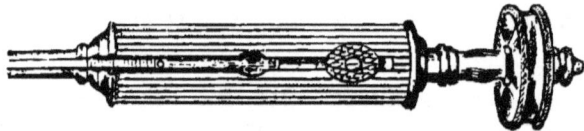

FIG. 17.—The handle of this lithotrite is shown; as it is this which affords the power to operate with great facility.

In what respect does it differ from others? In this, that it enables you to operate in less time, and with less movement or shock to the bladder, than any other instrument; and time, you know, is a matter of importance. It makes a good deal of difference to a patient, whether you retain in his bladder an instrument for three minutes or for one minute. If you pass a bougie into your own bladder at once and withdraw it, you may experience very little discomfort; but leave it there for three or four minutes, and see how you like it. Every half minute after the first the pain increases. The mere sojourn of the instrument in the bladder is a source of irritation precisely corresponding to the time, within certain limits, it continues there. Anything, therefore, that will diminish the time of the operation, and the amount of movement and concussion, will necessarily give a greater prospect of success.

Now for the proceeding itself. There is no operation that

I know which demands attention to so many minute details, all of them being very important. For its successful performance, it is essential that the surgeon should not only attend to the operation pure and simple, but to all the details connected with the case. Lithotrity neglected had better not be done at all. Either let it be done according to certain principles, and with great attention to detail, or let lithotomy be performed instead. One naturally recoils from many details; it is therefore essential to find out what principles regulate them. And, happily, these are very simple. What is the problem to be solved? It is the removal of a stone without injury to the bladder, either in employing the instruments, or by the action of the fragments themselves. That is what we have to aim at, and if accomplished success is certain.

Now, I need not say that by any cutting operation that is impossible. There is at the outset a severe injury to the patient in the shape of a large and deep wound, and it is this which, in any form, is always a risk. Let us see how far we can hope to solve the problem by lithotrity. All the chances of injury possible arise from these two sources: the stone itself, and the instruments used to extract it.

First, the stone. In its natural condition, as we know, it occasions no dangerous injury to the bladder, although it causes much pain, and ultimately chronic disease. But when it is broken up into large angular fragments with sharp edges, it becomes a source of injury, and severe cystitis may be thus readily induced. It is in accordance with my principle that I advise that these should be pulverized one at a time; you do not attack them all at once, and break them up indiscriminately into sharp pieces. Then you will take care that this débris is not hurried away when first made, and

I 2

while it is sharp; if it can remain in the bladder two or three days before it passes, it will become somewhat water-worn, and will pass more easily; at the same time the urethra will be less sore, and in better condition for its transit. It is as well also sometimes to promote the flow of urine by giving diluents and diuretics.

Secondly, the instrument used, and the method of manipu-lating it, may be productive of much injury, both to the bladder and to the urethra. It has therefore been an object with me to lessen, as much as possible, the number of instru-ments employed, the amount of manipulation applied to them, and the time devoted to the process. I have shown you how, in conformity with this principle, I have endeavoured to produce an instrument which should give the least possible irritation. And I will only add, that if we can get an instru-ment which will do its work with less disturbance still, it will be, *pro tanto*, a valuable step in advance.

Now as to diminishing the number of instruments used. Formerly it was laid down as an axiom that you should never use a lithotrite in a patient's bladder, unless it contained a known quantity of urine or other fluid. Hence the urine was always withdrawn before introducing the lithotrite, and four or five ounces of water were injected. I have shown that these preliminary injections are wholly unnecessary, and I never use them—never even asking a patient to hold his water beforehand, nor when he micturated last. It is said, " If there is only a small quantity of water, how can you be sure that you will not injure the coats of the bladder in endeavouring to seize the stone ?" There is no difficulty in that respect, because these instruments are so constructed, that you could scarcely lay hold of the bladder with them if you tried to do so. When you had instruments in which the blades closed upon each other accurately it was different. But these blades never do so : hence the safety of the instrument.

Then there is another species of mechanical irritation which may be met with. It was common—and it is now with some surgeons—to withdraw large fragments of stone from the bladder through the urethra. The surgeon would lay hold of them with the forceps, or with some other contrivance, and would really seem to think he had accomplished something to be proud of if he drew from the bladder a calculus as large as a bean. Now, in order to draw out such a calculus, you must first catch it. Well, if you have once caught it, why not give it one turn of the screw and reduce it to powder? Why subject the neck of the bladder and the urethra to pain and injury by dragging through a distance of six or seven inches, through a delicate and sensitive canal, a sharp angular fragment of stone ? So far from looking upon it as an achievement in surgery, I look upon it as a thing especially to be avoided. Never, then, on any pretence, withdraw an instrument containing a fragment or débris too large to pass easily through the urethra. Our object is mainly to crush the stone into small broken material, which will then pass harmlessly and easily enough. Again, it was the custom, after crushing, to inject repeatedly and forcibly several ounces of water into the bladder, the patient being placed upright, in order to remove the fragments just made. Now, this is an irritating process, more so often than the use of the lithotrite, and for the reason just named I regard it as a useless and meddlesome proceeding. You see, then, that we get rid of·the preliminary injections, the after injections, and also the withdrawing of the fragments. As a rule, everything is to be done with the lithotrite. One good flat-

bladed lithotrite will do seven-eighths of all the work. Other
means, Clover's apparatus for example, which is the best of
all, may be employed in exceptional cases.

Having enumerated what I believe to be the simple principles of lithotrity, I will further illustrate the practice.
Usually, when you have a large stone to deal with, it is
necessary to begin with the fenestrated instrument—that is,
one in which the female blade is entirely perforated, allowing
the male to pass through it.   This is always a more or less
dangerous instrument; hence it is used as little as possible
(Fig. 18). I never use it unless the stone is actually so large

FIG. 18.—The fenestrated lithotrite.

that it cannot be crushed by the flat-bladed instrument.
The edges of the instrument exactly meet, and are sharp, and
the fragments made by it are always rough and irritating.
Always, when it is possible, I use the lithotrite with flat blades
—blades which reduce the stone to powder (Figs. 19 and 20).
The blades do not meet each other, and cannot catch or
hurt the bladder, and the movement altogether is easier than
that of the other instrument. [A patient is brought in.]

I have told you that there is a difference in the mode
of introducing the lithotrite and the catheter. You know
that in passing a catheter, we, in this country, stand
on the left side of the patient; in France, the surgeon
stands on the right side.   In passing the catheter for a

recumbent patient, you hold it somewhat horizontally, draw
the penis gently over it, and give a gentle sweep, in this
way, into the bladder. In passing the lithotrite a different
movement is required. You may stand on either side, but

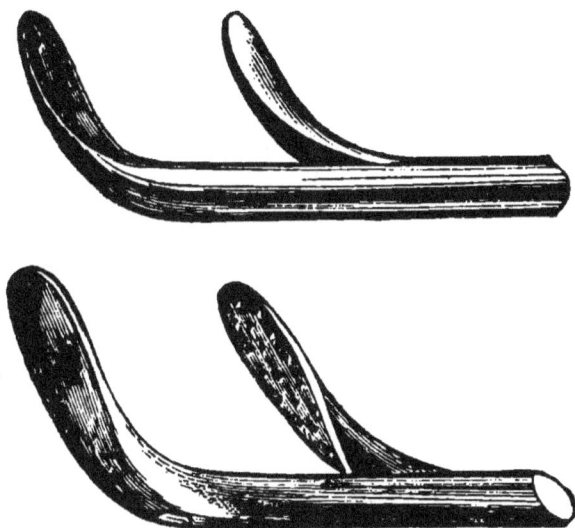

FIGS. 19 and 20.—Lithotrites with flat blades.

it is better to be on the right side, because that is the convenient side for operating, and it is awkward to go round
the patient to operate after passing the lithotrite. Well,
then, standing at his right side, and partially turning
your back to his face, you let the lithotrite slowly and easily
find its way until the shaft reaches nearly the vertical direction. Arrived at this point, you retain it in that position for
a few seconds, allowing it to go on, still so placed, by its
own weight, in order that it may slip under the pubic arch.
This done, you gently depress the handle, and it slides
readily into the bladder. There is no more easy instrument
to pass than the lithotrite with proper management. I have

now introduced the lithotrite, and have to find the stone and
seize it.   In order to do this I simply open the blades and
close; the stone is between them.   I touch the little button
here, which changes the sliding movement into a screwing
one, turn the handle and crush.   I then disengage the button, again open and close, and now I have a large fragment
between them, and, repeating the action, again crush.   A
good quantity of débris results; less than a minute has
been occupied, and I withdraw the lithotrite slowly and
gently, and here is some of the débris, which you see is uric
acid, between the blades.   There is no trace of blood, and
the patient has made no complaint of pain.   If you ask
him he will, I dare say, tell you it was not agreeable; but it
is nothing to take chloroform for—nothing like extracting
a tooth, for example.   Now we will have another patient
[who is placed on the table, the other walking away]; he
has had two sittings before, and knows all about it.   I introduce the instrument in the same way as before.   I open
and close the blades, and find nothing.   Turning them to
the right, I feel nothing; turning to the left, I feel nothing.
I then depress the shaft, and turn the instrument over, so
that the blades are reversed and point downwards; I open
and close, and seize a small fragment deep behind the prostate.   Having crushed that piece, I open and close in the
same position, and find one a little larger, and having
crushed it I withdraw the instrument.   This is always rather
more painful for the patient than when the stone is found in
the usual situation, and it occupies more time, perhaps two
minutes instead of one.   It is not usual to have a case of so
much difficulty, if you call it difficulty, as that.

Now let me give you a hint about crushing, which is a
very useful one.   Whenever you have found a stone, or a

good-sized fragment, and have crushed it, keep the lithotrite exactly in that place, and although you may have had
some trouble in finding it, you will now continue to find it
several times running. It reminds me of fishing for perch ;
when you have caught one, you may catch, perhaps, twenty
or thirty more out of the same hole, if you will but stop
there, and not go fishing about among the shallows. It is
the same in lithotrity. You will go on seizing and crushing
if you contrive to keep the lithotrite precisely in the same
place. In fact, there is what may be called a certain favourite
" area " in every bladder in which to operate—a certain spot
which is a favourite haunt, so to speak, for fragments of
stone. If you find that out in each bladder you will always
be able to crush ; if you do not, you may often have some
difficulty in discovering your stone. The area will, of
course, vary somewhat with the position of the patient. If
the patient was standing, for instance, the area would not be
the same as in a lying posture. It is best to raise the pelvis
two or three inches ; then you get an area for operating
which is not too close to the neck of the bladder. The neck
of the bladder is a very sensitive part, and you should always
avoid it, because in pulling out the male blade you may im-
pinge against the neck of the bladder if you are not careful.
One of your maxims in lithotrity should be never to pull
out forcibly the male blade. You should pull it out carefully and delicately, so as to *feel* the neck ; and it is a bad
lithotrite, remember, if the male blade does not slide with
perfect ease. This diagram will show you what I mean by
the area for operating. If the patient is lying without a
cushion, it will be nearer the neck of the bladder than if the
pelvis is well raised. In the last case you saw just now, the
man had an enlarged prostate, and when the lithotrite was

introduced the stone was not found, as you saw, until the
lithotrite was reversed.   It is more essential in the case of
an enlarged prostate to put a high cushion under the pelvis,
in order to throw back the fragments to the posterior part of
the bladder, so that the area may be as far from the neck as
possible.

There may be some advantage in operating on dead
bodies.   But there is a great difference between the feel of
the thing in the living and in the dead body.   The conditions encountered in the mere flaccid bag which constitutes
the bladder in a dead body is very different to that in the
living.   If you have a master to give you lessons, it may
serve your purpose, but not nearly so well as in the living
body.

So much for the sittings.   But it is of great importance
to watch well the patient afterwards.   You heard me say in
the theatre the other day, and I may shortly advert to it
here, that you should not encourage the early passing of the
fragments.   They rest at the bottom of the bladder; and I
usually keep the patient in bed, and pretty much on his back,
for thirty-six hours or so afterwards : he should, at all events
for that period of time, pass urine in that position, so that
the sharp angular fragments are left at the bottom of the
bladder, and are not forced into the urethra.   This used to be,
and still is, much neglected, and the consequence is, that if
a man has an urgent want to pass water, he will get up and
strain.   The sharp fragments are driven into the neck of the
bladder by the effort, and bleeding, pain, sometimes inflammation of the prostate or of the testicle follow.   But
by adopting the position named, the sharp fragments get
waterworn, and the urethra, which has been irritated, and is
perhaps a little swollen from the contact of the instrument,

recovers itself, and the passage and fragments are better adapted to each other than before. I believe this is not taught anywhere but by myself; it is a very simple matter, but an important one. Never allow a man who has been subjected to lithotrity to micturate in the standing position between the first and the second sitting, when the fragments are all large and sharp. When you have crushed two or three times, perhaps, he may get up and make water occasionally ; but after each sitting, for twenty-four or thirty-six hours, he should not pass water in the erect posture.

Immediately after a sitting, a hot linseed poultice may be placed above the pubes, and is a comfort to the patient. The bowels should act the day before or on the morning of the sitting, so that the patient may not have to rise soon after and strain.

Now, suppose that we have—at five, or six, or eight sittings, according to the size of the stone—almost if not quite removed it, a very important duty remains. It has been objected to lithotrity—and there was some truth, perhaps, in the allegation formerly, but not now, if the operation is done well—that you never make sure of getting rid of the last fragment; that you might leave a portion to become the nucleus of a future stone. But there is very little more difficulty in getting away the last fragment than any other, provided you go the right way to work. Generally speaking, in four cases out of five, the last fragment passes as the others—that is, by the patient's natural efforts. But supposing there remains a bit too big to pass—suppose you have reason to believe this from the continuance of the pain, &c.—you then take an instrument with short wide rounded blades, with which you can explore easily in the reversed position. With such a one you may thus search the

whole floor of the bladder with perfect safety. Now, supposing
you have done this and have found a bit of stone, and do
not know whether there is another—some little symptoms
are left, but you do not find any more—what is to be done?
You do not know whether these symptoms are due to a fragment of stone which escapes you, or whether to the irritation
arising from the long residence of the stone in the bladder,
and also, to a certain extent, to your instrumental efforts to
find it. This is what I advise: wait for a week and see
whether the patient is better, and if there is any doubt at all
about it, make him take some severe exercise. I do not
know anything here in London better than a long omnibus
drive. Tell him to expend a shilling in omnibuses—say
between Mile End and Kensington,—and if that does not find
out the fragment in a man's bladder, I scarcely know what
will. If he can stand that, especially if the roads are mend-
ing—if he is not worse for such a journey,—rely upon it
he has not any fragment left: the irritation is due to the
state of the bladder, and not to a fragment. If the latter
exists, the patient will to a certainty be the worse for his
drive, and perhaps there may be some bleeding. I admit
that it is difficult to tell sometimes what the irritation is due
to ; and it is only by waiting and testing in some way that
you can find out.

Then there is another very good mode of getting rid of
the last fragment, and also of removing fragments at any
time when they are not well expelled by the natural efforts,—
I mean the apparatus of Mr. Clover, who, well known by his
chloroform instrument, has also given us a valuable addition
to the instruments for lithotrity. It is a very fascinating apparatus, and looks so well that there is some risk of its being
used when not required. There are certain conditions in

which the fragments never come away well. For example, a patient has no power of making water, except by a catheter; then few fragments, probably, will be passed in that way, and you will require other means. For him the apparatus in question is the very thing. After crushing, this large catheter is introduced, and the fragments are sucked out by the action of this powerful india-rubber bottle which is attached to it. The process is rather trying, however, for the bladder; and it costs rather more pain and time than an ordinary sitting for lithotrity.

One troublesome thing that very rarely happens is the impaction of a fragment. If you adopt the system of lithotrity which I have shown you, it is remarkable how seldom that happens. I have never had to open the urethra to remove a fragment in my life. I have occasionally had to remove one by the forceps, but that is very rare; and among all the complicated inventions for the purpose, I know nothing so good as the common long forceps which I show you here. During the last year, certainly, I have not even had occasion to use them. The more thoroughly you crush the stone, the less use there will be for forceps. Here is a bottle containing what I call a well-broken stone. You see it is almost powder: a very different sort of thing from that in the other bottle, where you see a large number of big fragments that were probably passed with difficulty. It is an old saying, " a carpenter may be known by his chips : " certainly the skill of the lithotritist may be known by the débris he makes.

One other point I have to mention—namely, should you employ chloroform? Some operators prefer it. Now, first of all, I think the operation ought not to be sufficiently painful to make chloroform necessary. If it was like extracting a tooth, I should give chloroform. But there is a better reason than that for not giving chloroform : without it you can better judge of the susceptibilities of the patient at the time. On some occasions the bladder is more irritable than on others ; and if the patient suffers much, you do not have so long a sitting. If he scarcely suffers at all, you go on " making hay while the sun shines," and crush three or four times. On ordinary occasions, two separate introductions of the lithotrite suffice. Too much lithotrity, when the bladder is irritable and painful, is usually an equivalent to doing the patient some mischief. This is a point as to which you cannot judge, if chloroform is administered. It has been said that chloroform should not be given, because if the patient is not under its influence, he will call out if you seize the coats of the bladder in mistake for the stone. That is altogether an error. The coats of the bladder should never be injured, and with proper instruments it can scarcely happen.

I had intended to speak to you of certain contingencies that may happen, but it is now too late to dwell upon them. I will simply name them ; they will come into consideration in their appropriate places in subsequent lectures. The first contingency is fever ; secondly, bleeding ; thirdly, cystitis ; fourthly, orchitis ; fifthly, retention of the urine ; and lastly, exhaustion, which is sometimes fatal. A peculiar kind of feverish attack is, as you know, common after all instrumental operations on the urethra. The phenomena of cold chill, dry burning heat, and sweating, proceeding in this order more or less severely. Do not " meet these symptoms" by a too active treatment ; I know none of much service ; when the patient is thirsty, give him drink, and don't press food on him until he is somewhat disposed for it. Rely on it, what we call fever here, is nature's struggle against some poison in the course of elimination ; only take care that the hygienic conditions are good. After the attack he is weak, and requires good and nourishing food.

Bleeding, as a result of the operation, is very rarely troublesome, and does not require much treatment. Cystitis gives a little trouble occasionally, and is to be treated in the ordinary way described in our course. Inflammation of the testicle requires you to desist for a time from operating. Chronic retention of urine is apt to occur very insidiously— not absolute, but partial retention. Always look out for it, if frequency of micturition increases, and the urine becomes increasingly thick, and if the bladder is not emptying itself, you must pass the catheter once or twice a day. Sometimes exhaustion occurs ; after a number of sittings the patient's strength gives way. That, however, is very rare.

In my next lecture I shall take the subject of lithotomy, and give you a general sketch of the different modes which have been practised, and which are being used at the present day.

LECTURE IX.

LITHOTOMY.

GENTLEMEN,—It was understood, as the result of our discussion in the first lecture on Stone in the Bladder, that all
cases below the age of puberty, with very few exceptions,
should be cut; those only in which the stone is quite small
being reserved for lithotrity.   Then there were certain adult
cases, where the stone is large, or other difficulties exist, in
which the operation of cutting must be performed.   Hence
we have next to study the operation of lithotomy, and this
advantageously follows the consideration of lithotrity.

The proceeding of "cutting for the stone" has always
been a subject of extreme interest; indeed, you will find
no operation that has exercised a greater fascination for the
veteran operator, while there is none which more excites
the ambition of young surgeons.   There is none respecting
which an old student comes to his teacher here with so much
just pride and happy sense of newly acquired power, as when
he tells him, "1 have just cut successfully my first stone
case down in the country."   On the other hand, the true
surgeon who loves his art, is always at home when the theme
of discussion embraces the history and practice of lithotomy;
and 1 verily believe that some acquaintance with the former
is one of the best ways of commencing a study of the latter.
1 will therefore give you a sketch, although it must be very
slight and meagre, of the different stages by which lithotomy
has arrived through early periods at its present condition.

The first record we have of lithotomy is in the Augustan age. I shall speak first of lithotomy under the Classic period, then under the Mediæval period, and then under our own period—the period in which, without regard to what has gone before, and with the utmost indifference to authority or antiquity, we sift anew everything to the bottom, in what we call "the spirit of the nineteenth century." I dare say there was an earlier period still, even before the time of Celsus, the author of the first account (who, by the way, was a physician, and never operated) ; and if some enterprising Lyell in surgery would make the inquiry, he might find traces of a prehistoric period ; because wherever there are human remains, calculi must exist. I do not know that uric-acid stones would endure long. We know that the excreta of fishes are preserved for thousands and thousands of years, and I doubt not that some of these human excreta might be found also, and that oxalate-of-lime calculi, at least, must exist, among other human remains. As so many observers are seeking the early records of the human race, I throw out the hint ; and certainly, if I were so searching, I should not forget to seek, among other things, the matters in question. Whether we shall thus ever find any instruments which could be identified as the means by which those stones were removed is doubtful. We will, however, not occupy our time with speculation, but will be satisfied to begin with such facts as we can find, say about 2,500 years ago. More than 2,000 years have passed since Celsus wrote, and hence I have named that date, as the operation was evidently an established practice at that time. Hippocrates speaks of it, although he made his pupils take an oath that they would never practise lithotomy. For then lithotomy was an occupation by itself, and not a very exalted one in popular esti-

K

mation, being only practised by certain itinerant performers. What they did was described by Celsus in his seventh book, and it was called "cutting on the gripe." The method was simple, and so were the instruments, on which account they were afterwards termed the "apparatus minor," to distinguish them from the "apparatus major," which came into vogue in the Mediæval period. The operator commenced by placing his patient, usually a boy, upon the knees of a man who was seated. If it was an adult patient (but such were rarely cut), two men sat side by side (their legs forming the operating table), so that their arms might clasp the patient and control his struggles. The operator used no staff whatever, but inserted two or three fingers into the rectum, and endeavoured so to feel the stone, which he could only do if it was large. If he succeeded in this, he fixed it firmly, or "griped" it with the ends of his fingers—hence the term "cutting on the gripe;" and pressing it down towards the perineum, he made a semilunar cut with a broad scalpel until he reached it. Then, if unable to press it out with his fingers, he drew it out with a hook. Now this rough-and-ready mode was extant until about the sixteenth century; indeed, up to the seventeenth century it was largely practised in Europe. Even in the latter part of the seventeenth century, when Frère Jacques appeared, the ancient mode of cutting on the gripe was chiefly practised.

We now come to the Mediæval period, when at least three different operations appeared. Appropriately enough, too, a monk figures now as the most famous operator.

First, we will consider the "Marian method," or "apparatus major"—a median operation, originated by Johannes de Romanis, but receiving its name from his pupil, Marianus Sanctus. It is called the "apparatus major" because, while "cutting on the gripe" required only a knife and a hook, this table would be scarcely large enough for the instruments employed for the Marian operation. They are not here, but you may see them at the College of Surgeons. By this method a vertical incision was made by the side of the raphé, and the urethra was opened on a staff at about the membranous portion. A dilator was then thrust into the wound, and another upon that (male and female dilators they were called), and the canal and the neck of the bladder were torn asunder with great rudeness. Its only resemblance to the present median operation is, that the incision is in nearly the same place. But anything more barbarous than the practice you can hardly conceive. The stones were larger then than they are now, and the incision was small; and in order to dilate and extract, various machines were employed, the result of much mechanical ingenuity, and forming the origin of half the surgical instruments now in use. It was a very unsuccessful operation, and was gradually lost sight of in consequence. Still it held its ground in places, and for certain cases, as late as to a part of the eighteenth century.

Next I shall name the high or suprapubic operation, which appeared about the end of the sixteenth century, and this has maintained a position of greater or less importance to the present day. More than this I shall not say on the present occasion, as we concern ourselves mainly with perineal lithotomy.

I now come to a new proceeding, which rudely shadowed forth our present lateral operation. It was performed on a staff, which was not grooved as now, but yet it roughly served as a guide into the bladder. The operator commenced by thrusting a long knife into the ischio-rectal fossa,

K 2

and so on into the bladder behind the prostate, and, cutting forward, he made the entire wound at one incision. Invented, as it was believed, by Pierre Franco (sixteenth century), its apostle and promulgator was the celebrated Frère Jacques, who flourished in the seventeenth century, and is said to have cut 5,000 times for stone. Probably he did not cut 500 ; but a cipher more or less was a trifle for the inexact and credulous mind of the period. Like others of his craft, he was an itinerant operator, not embarrassed with too much anatomy, and France was mainly the scene of his labours. After him, Rau, in Holland, pursued the same practice.

It will be interesting to you to know what was going on meantime in our own country. Most patients, up to the end of the seventeenth century, who were cut, submitted either to the old operation "on the gripe," or to the "Marian." In the beginning of the eighteenth century, the high operation was first practised here. At this period there came to London a Leicestershire lad, subsequently known as Cheselden, the celebrated surgeon of St. Thomas's Hospital; and he at first did the high operation. But he had heard of the recent successes of Frère Jacques' method, and tried it, modifying it as his experience suggested, until he performed almost exactly what we now call the lateral operation, and with the best results. His success was so great, that in 1729, when he had performed the operation several years, and cut about one hundred patients, Morand, the French surgeon, was sent from Paris to see him operate, and report upon the subject. He remained here for some time, Cheselden getting together a number of cases and operating on them before him. Morand then returned and reported to the French Academy so favourably on the subject, that Cheselden's

operation became generally accepted as the best. In it the
deep incision was made, if possible, within the limits of the
prostate gland, and involving its left side only, by means
of a scalpel of moderate size, cutting its way inwards along
the groove of the staff. A few years afterwards Cheselden
retired, having cut 213 patients of all ages, with ten deaths.
Those are the first figures that we can depend upon in connection with the operation, for, as I have told you before, the
figures of the Mediæval period are monstrous and incredible ;
for not only was the famous monk said to have cut 5,000,
but to have lost " scarcely any." Cheselden, who improved
the mode, and cut at all ages, including very many children
(who are, as you know, extremely safe to cut), had barely
five per cent. of deaths, which was an enormous result, and
no doubt the greatest that had been yet made.

At this point the operation continued for some years until
the end of the century, and then " the gorget " came into
fashion. A few years ago a patient was rarely cut without
it ; now, I suppose, few of you know what it is. Originally,
one of the directors used in the " apparatus major," its edges
were sharpened for the purpose of making the deep incision through the prostate. This was the idea of Sir Cæsar
Hawkins, whose name was affixed to it ; but subsequently
almost every surgeon had his own gorget, making it wider
or narrower, or altering it in some fashion or another. A
great deal of mystery has been made of this instrument, but
it is simply a wide knife with a beak at the end, which is
kept in the groove of the staff. In using the ordinary
knife, if you require a deep incision, the blade must leave
the staff a little. The object of the gorget is to make an
incision sufficiently deep without leaving the staff. Here
is one which formerly belonged to Scarpa, the celebrated

anatomist, and here are others which, having been used by many celebrated operators, have fallen into my hands, and they are leading types of the instrument.

In 1816, Dupuytren, of Paris, not being content with the lateral, introduced his bilateral operation. His object was to make the deep incision by a cut on each side of the prostate, instead of one large one on one side only. And he, with the same view of limiting accurately the extent of the wound, designed a special instrument for the purpose. This, the " two-bladed lithotome," is also a member of the ancient surgical armamentarium, made more elegant and manageable by modern skill. Instead of making the internal incision by pushing a cutting blade *inwards*, as with the knife or gorget, you carry this instrument [showing it] into the bladder along the staff, there open its two concealed blades, and drawing it towards you, cut your way *outwards*. You can arrange the blades so as to have the incision as wide or as narrow as you please.

In 1825 or 1830, the "median operation," often loosely spoken of as a revived Marian operation, came into some note in this country, and with it Mr. Allarton's name is chiefly associated. In the meantime Civiale, in Paris, combined the median and bilateral operations, and you have seen me frequently perform both of them here. Then Nélaton has recently performed the pre-rectal operation, which is, after all, only a carefully dissected bilateral operation.

I now come to the mode of performing the lateral and the medio-bilateral operations, and will give a few general hints which will apply equally to either. As I have said before, when we have to do with many details, let us try to revert to first principles, and define clearly the object we aim at. I told you that the object of lithotrity is to remove the stone without injury to the patient, either from the stone or the instrument. In lithotomy you must have a wound, and the object is to make it in such parts as shall least endanger the blood-vessels, the viscera, or the neck of the bladder, and to remove the stone through the lower outlet of the pelvis with as little mischief as possible to any of those parts. When that problem is best solved, we shall have the best form of lithotomy. It is quite open to discussion whether we have yet found out the best way, although we have been 2,500 years—to say nothing of the pre-historic period—in coming to our present position. Only the other day, Sir William Fergusson was proposing to revert in some fashion to the ancient classic form which bears the name of Celsus. Then Mr. Erichsen last year performed Dupuytren's bilateral method, which has become obsolete in Paris. And in that great school of surgery two or three kinds of operation are in vogue, and there is by no means an agreement as to which is the best.

Now, in order to aid you to solve the problem for yourselves, I have placed before you a diagram drawn accurately from the preparation, showing the bones and ligaments of the pelvis, in the position for lithotomy. The lower outlet is opposite to us; it is in the patient filled by soft parts, and it is the opening into which you have to cut, and through which you must remove the stone, and in all that you do, you must be limited by its boundaries of bone. I like to have that in my mind's eye when the patient is tied up and I take my seat to operate. Here also are diagrams, showing two stages of the dissection of the perineum. But I take it for granted that you know your anatomy too well to require any detailed account here of the important parts involved in the operation. I shall simply name those

which concern us. First, there is the pudic artery, safely
sheltered under the pubic ramus; but it gives a branch to the
bulb, a vessel to be avoided at the upper part of the space.
Then in the same part is the bulb of the urethra, which is
not to be thought too lightly of; indeed, it is the source of
some of the chief dangers; it is a vascular expansion from
the vessel named, and cutting into it deeply is as bad as
cutting into the vessel itself, if not worse. Next, there
is the rectum in the middle and lower part, which it is
also important to avoid. The other diagram shows the
position of the prostate, which must be divided in the deep
incision.

I will now very briefly touch on the principal steps of
the operation. The patient's bowels are to be thoroughly
emptied by an enema a few hours before. Do not trouble
yourself about the quantity of urine in the bladder. Some
think it very important that it should be full. Cheselden,
on the other hand, preferred it to be empty, saying that in
this condition the stone was easily found close to the neck
of the bladder. I have seen great pains taken to inject the
bladder before operating; but the unconscious patient has
usually succeeded in emptying it, in spite of tying the penis,
and such-like precautions.

The first thing the operator does is to pass the staff into
the bladder and find the stone. Never think of cutting a
man if you are not fully satisfied that the staff is in contact with it. Frightful blunders have been made through
indifference to this rule. Suppose, for example, the staff is
in a false passage, and is not in the bladder at all; one
shudders at the idea of an operation performed on a staff
so placed—an exhibition distressing to all concerned—never
to be forgotten either by the operator or by the bystander, and

probably fatal to the patient. The "click," then, is to bo distinctly audible to yourself and to a witness, and the staff is to be put into the hand of your best friend, who is to attend implicitly to your instructions, and to no others, whatever they may be. The patient is then to be tied up firmly; better still if secured by these leather anklets and wristbands, devised by Mr. Prichard, of Bristol, because they truly realize the proverb, "fast bind, safe find," which our old friends the garters often did not.

Now, what are the instructions to your friend, the staff-holder? You want it held firmly, and, of all things, not to leave the bladder. I don't think you will gain much by cultivating a fancy for any particular spot, such as right or left, or projecting in the perineum. If it is to be steady and in one spot, which is the main thing, there must be a point of support for it to rest against, and there is but one such spot in the whole region. Rely upon it, then, you had better tell him to keep it close to the arch of the pubes, well hooked up, with the handle pretty nearly vertical. Your fingers now traverse the region and find the lines of the rami, also the condition of the bowel, whether empty and contracted, or the reverse.

Now, relative to the first incision, different authorities advise different places at which to enter the scalpel and commence. Without discussing these, let me say that, as a rule, the usual spot should be, for an adult, about an inch and a quarter in front of the anus, a little on the left side of the raphé. Go in boldly, slightly pointing upwards, near to or into the staff, and then gradually less deeply till you come out about three inches lower down towards the inner side of the tuber ischii. It is very pleasant to feel that you touch the staff in that first incision, and it saves trouble and uncertainty to have gone close to it, which you always ought to do; never let it be a timid shallow cut, merely dividing the skin. The end of your index finger follows, and should feel the staff easily through the tissues. Fixing the finger-nail upon it, the point of the knife is placed firmly in the groove, and is run steadily on in contact with the staff. Keep the point up, and you will be safe; let it down, and you may slip out and get into the rectum, or nobody knows where. Simply go on, letting the blade be a little more horizontal as it proceeds. Go on till you are well into the bladder, not letting the point leave the staff. The depth of the incision will depend upon the angle which the knife makes with the staff: if you withdraw with the knife close to the staff, of course you will only make a wound the width of the knife; and if the edge is directed outwards and downwards against the soft parts, with a light hand, as you come out, you will make a freer and cleaner opening. It is better to be rather free in cutting than otherwise [the presence of a large stone is assumed], but you must not make the incision too wide. There has been a great deal of good advice expended about this subject—the depth of the incision, but it is manifestly impossible for one man to make another understand what he means or what he does by any amount of talk. My belief is, however, that the result of our anxious care about this matter is, practically, that we are apt to cut rather too niggardly than too freely, and that the neck of the bladder, in consequence, receives severer injury from the stone and forceps than it otherwise would receive from the knife. This relates, of course, to adults; for in children you can scarcely find the prostate—it weighs but a few grains, and does not come in for a moment's consideration, and your knife goes far beyond its limits; yet these little patients are the safest to cut. Of course there is a very great difference in the two ages, due to the different conditions of puberty and childhood. To return. The incision being completed, your left index finger immediately follows close along the staff into the bladder, where you will probably just touch the stone. The finger goes firmly and deeply in, stopping the urine perhaps to some.extent in its outflow, and accomplishes some dilatation of the parts. Then you slide the forceps closely along the palmar surface of the finger, and insinuate them into the bladder, which makes dilatation number two. Then, generally speaking, you have but to open the instrument carefully, yet widely, one blade flat at the bottom of the bladder, the other towards the top, and, closing the blades, the stone is probably between them. If it seems that you have a good hold, draw gradually outwards and downwards, easing or adjusting, if you can, with the left index; and so you make the third and last dilatation. Remember not to pull out horizontally and bruise the parts against the pubic arch, but downwards into the widest part of the lower pelvic aperture. And don't be hurried for the sake of anybody else. You and your patient are to be, for you, at this moment, the only persons present, and your responsibility to him must never be forgotten for an instant through the influence of bystanders and lookers-on.

I must now briefly add that you will search for a second stone, tie any vessel spouting within sight, insert the tube into the bladder, perhaps inject a syringeful or two of cold water, and stuff the wound round it if the hæmorrhage is free. The patient is placed in bed on his back, with one or two pillows under each ham, and the parts involved exposed to air and light, so that you see how the urine flows. The less meddling afterwards generally the better.

I have only time to say a word or two about the median and medio-bilateral operations. For the median, an incision is made in the line of the raphé from about two inches and a half above the anus, downwards as near to its margin as is safe, for you want all the space you can get. Dissecting down to the staff, with a finger in the rectum, and opening the urethra in the membranous portion or thereabout, you carry a director on into the bladder; your finger follows, and dilates, and then the forceps on that. I may tell you that it is generally performed by thrusting a straight bistury, back downwards, in front of the anus, into the staff, and cutting upwards and outwards at one incision. I prefer the other mode. Manifestly this operation will not do for large stones, which mainly, thanks to lithotrity, are what we have to deal with now. Hence the applicability of the " median " is extended by making it " medio-bilateral," and in this manner: Having performed the median, as just described, up to the point of opening the urethra, instead of intro-ducing a director, you introduce the two-bladed lithotome, and when it is in the bladder you open the blades, and two moderate incisions are made, one right, the other left, as you draw the opened instrument outwards in the groove of the staff. These two operations I have now performed about thirty times, and I do not know, after all, that there is much to choose between them and the old " lateral." To make an accurate estimate, one requires at least 100 cases of each operation by the same hand. Nevertheless, I may say a word, finally, on the principle which essentially distinguishes these operations. They are the result of opposite convictions respecting the hazard of the knife. There is a set of men to whom anatomy is a bugbear, and who are afraid of cutting as much as is absolutely necessary; and there are other men less timid,—mind, I don't say less cautious,—who regard the larger freer style of operating as better than the small or fearful style. All surgeons, of course, tend more or less to fall into one of these two classes. The anatomical school have devised a variety of median operations in order to avoid certain blood-vessels, &c., and they sacrifice space in doing so. They answer excellently well for small and medium-sized stones; but these are, or should be, crushed now, and we do not want any operation for such stones. The perineal operation which offers the most room, the recto-vesical excepted, is the lateral operation. All the others named are essentially median operations. Now, I am bound to say that formerly, judging theoretically, and performing less lithotrity than now, I had a leaning to median methods, being disposed to think that they would be attended with less hæmorrhage than the others. But I do not find this so in practice, and I have arrived at the conclusion that there is quite as much bleeding as in the lateral operation. I attribute this to the bulb. I regard the bulb as a large artery to all intents and purposes. You cut into that spongy tissue,—not in all cases, but in some,—and there is as much bleeding as if you cut the artery of the bulb, and more difficulty in controlling it. The bulb must be cut more or less in the median operation. The problem is how to get into the bladder without wounding the bulb, its artery, and the rectum; and I believe that a well-performed lateral operation accomplishes this, where a free opening for a large stone is required, better than any other.

I cannot tell you which of these two operations is the easier; if anything, perhaps, the lateral. But here is the important fact, which is but beginning to be realized by the profession, viz. that only exceptional cases of large stone in the adult require any cutting operation, since all the small and middle-sized ones can be much more safely removed by the crushing process; and it is this fact which is bringing these various forms of lithotomy under the full and serious consideration of surgeons at the present day.

The next lecture will relate to diseases affecting the bladder itself.

LECTURE X.

CYSTITIS AND PROSTATITIS.

GENTLEMEN,—It is quite possible that you may see little or nothing, even during a considerable experience of general practice, of those diseases which have occupied our last three lectures. Some men never meet with a case of stone in the bladder during a lifetime, and of those who do, very few undertake themselves to treat it. But the very reverse of this is the case with the subject which comes before us to-day. Fraught with much less of what one calls "interest" for an operator, its attractiveness to the student must be found in the fact that cystitis, or inflammation of the bladder, is the commonest affection of that organ, that it is certain to occur in his practice, and probably not unfrequently. For, whatever else you may have in connection with the urinary organs, you are certain to have cystitis. If a man has stricture severely, or stone, or disease of the kidney, or disease of the prostate, sooner or later he has cystitis, either acute or chronic, the symptoms of which are often the most prominent features in all these diseases.

Then I beg you to remember that cystitis has almost always some ascertainable cause, and that it very rarely indeed appears in what is called an idiopathic form. You will find that there is, or has been, gonorrhœa, or stricture, or disease of the prostate, or retained urine, or urine altered in character, or some other such cause ; and if you have readily come to the conclusion that any case before you is idiopathic, suspect that you have not discovered the cause, and the probability is that you have not searched deeply or carefully enough. Here and there the true pathology will elude our best efforts. Possibly you may be forced to attribute it to a gouty diathesis. A very refuge in time of trouble for practitioners of feeble diagnostic power is gout, particularly "suppressed gout;" therefore beware of it. And while, I think, it must be admitted that inflammations, both of the urethra and of the bladder, may be sometimes mere local developments of the ubiquitous influence so named, I am sure that this cause is of exceedingly rare occurrence. Certain irritant poisons also—amongst which cantharides is the most prominent and likely to be met with—occasion cystitis, which I have seen severe in character, and lasting from ten to twenty hours, as the effect of an ordinary blister.

First, then, I shall call your attention to acute, and secondly, and chiefly, to chronic cystitis.

Acute cystitis appears in two very distinct forms: one which is severe and dangerous, the other which is much less so.

The dangerous form is that which accompanies the most severe lesions to which the bladder is exposed. The first breaking up of a large and hard stone into large fragments, either spontaneously or by the lithotrite, has sometimes caused it; violence to the bladder in lithotomy and the like are causes. Rigors, bloody urine, extreme pain and irritability of the organ, announce the fact, and the patient succumbs in a few days. At the autopsy you will see the lining membrane of a dark crimson hue throughout, or nearly so, and often spots where it appears to be sloughing and exposing the muscular fibres.

The less severe form of acute cystitis is common enough. There is reason to believe that the neck of the bladder is the part mainly affected in these cases ; and the reason for this is that, what we call cystitis is often really inflammation of the prostate mainly, or of the urethra passing through it, entirely so perhaps at first, and that the mucous membrane of the bladder is affected by extension. And as, anatomically, I don't know how we can make an accurate separation between these two organs, it is often legitimate to speak of this affection as "inflammation of the neck of the bladder." After gonorrhœa, or from external cold and damp, and in connection with many circumstances of no apparent importance, a man becomes the subject of frequent and painful micturition, and has a sense of aching or gnawing pain above the pubes ; while the urine is cloudy from an increase of the natural mucus of the bladder (not the tenacious viscid secretion, mind, which I shall speak of under the head of Chronic Cystitis) ; and there is usually some constitutional sympathy with the local disturbance, evinced by loss of appetite and general feverishness.

The treatment consists in mild laxatives and diuretics, demulcents, hot hip-baths, local poultices, and anodynes, if necessary. Besides these, the use of certain infusions, and decoctions, of which I shall speak hereafter.

But that which most requires our attention is the chronic form of the disease. It is that which requires the most care and judgment, and for which we have most to do in the shape of treatment. Chronic cystitis also appears in two distinct forms. In the simpler, there is little else than some increase of the natural mucus from the bladder mixed with the urine. Just as when you have a common cold, there is inflammation of the mucous membrane of the nose and towards the frontal

L

sinuses, with increase of its secretion, so the mucous membrane of the bladder adds mucus to the urine; and the
inflamed mucous membrane, being more sensitive, will not
permit itself to be much extended by accumulating urine,
but forces the bladder to expel it as soon as possible: hence
the frequency of making water. But, besides this, there is
another form, in which the mucus has a distinct character.
It is often spoken of, and not very wisely, as " catarrh of the
bladder"—another example of an unfortunate term leading
to error in practice. The mucus is very tenacious, and when
you empty a vessel containing the urine of such a patient, it
runs off first, and then a quantity of ropy mucus follows in
a mass. You may see a pint or more of this material passed
in the course of the day, and it acquires the viscid character
on standing. Some patients pass it for months together,
and such are said, especially abroad, to have " catarrh of the
bladder," which, and there especially, is said to be a very
dangerous and a very incurable disease. Indeed, to tell a
foreigner that he has " catarrh of the bladder," is to alarm
the poor man exceedingly. Now, this is because it is
regarded as an essential disease, instead of a mere symptom;
for of course it is no more a disease than dropsy is. Formerly, you know, we talked of dropsy as a very formidable
malady, and it still is so to the popular mind; but no intelligent student would now, I suppose, be satisfied to think of
it except as a symptom. He would say: " Is it due to car-
diac, to renal, or to hepatic disease? What is the cause of
it?" Precisely so is it with this " catarrh of the bladder."
You inquire what is the cause of it, and you find in nine
cases out of ten there is a very distinct cause, and mostly a
removable one. You are not to be stopped by this name,
and ask me what is good for catarrh; but you must carry

further the diagnosis, and ascertain the precise condition which has occasioned it. And the most common cause is one too often overlooked, as I shall repeat—viz. inability of the bladder, either from atony of its coats or prostatic obstruction, to empty its coats; yet this peculiar muco-purulent secretion does by no means necessarily occur in these circumstances, and I cannot tell you at present how it happens that in some cases of retained urine it contains only some light flocculent mucus, and that in others there shall be a very large quantity of this tenacious matter.

With regard to the treatment, the first thing is to take care that the bladder is emptied by a catheter once, twice, or three times a day, in the easiest manner possible, of which I treated in the fourth lecture. And this is necessary because decomposing urine is a great source of irritation to the mucous membrane. The urea contained in the secretion which enters by the ureters in a healthy state is soon de-composed into carbonate of ammonia, and the ammoniacal salt is an acrid and irritating substance. You explain to your patient that his bladder, not having been emptied for many months perhaps, has acquired somewhat the condition which a badly washed utensil would have done in like circum-stances—a useful and sufficiently accurate illustration for the lay understanding, and he will appreciate it when he finds, as he probably will, that the mucus diminishes con-siderably after a few days of this treatment. But suppose it does not do so, or does so but slightly, what then? I will tell you what sometimes happens, and I am not sure that the fact I am about to ask your attention to has been observed or recorded. It is this: *you cannot completely empty every bladder with the catheter*. When the prostate is irregular in shape, and throws out protuberances into the

L 2

bladder, there are sinuses or spaces between them, which retain one, two, or even more drachms of urine. Again, there are not unfrequently numerous small sacculi in the coats of the bladder which act in the same way. When obstruction at the neck has existed some time, the daily straining—although not considerable—necessary to expel the urine, produces hypertrophy of the muscular bands which form that coat of the bladder. Now, you know hydraulic pressure is equal in every direction, and in course of time the expulsive act, more powerful than in health, gradually forces the mucous lining between the interlacing muscular bands, and little pouches result. In these it is not uncommon for calculi to secrete themselves, and thus in time encysted calculus is formed. In any case, however, those pouches become receptacles for urine, which becomes stale, and irritating, in consequence, and not at all unfrequently they attain a large size ; such a one is depicted at Fig. 21. Now, the mere withdrawal of the urine by catheter by no means empties the reservoir in these circumstances, and enough of noxious fluid is left in these pouches or sacculi to maintain the unhealthy condition of the lining membrane. What you have to do is, to wash out the bladder at least once a day with a little warm water before you remove the catheter. I am very particular indeed as to the manner of doing this. Washing out the bladder may be a very valuable mode of treatment, or a mere contrivance for seriously irritating that organ, according to the mode in which it is performed. A common mode—indeed that which I always saw employed some years ago—was to attach to the catheter (which was often of silver, and it is unnecessary to repeat my views about that) a large metal syringe, and to throw in with considerable force six or eight ounces of water. I wish you to cherish a wholesome horror of that proceeding, and in no case can it be necessary. A healthy bladder, and much more, a tender one, can only be disturbed and pained by such a process. This sensitive organ is only accustomed to be

FIG. 21.—Section of bladder and prostate. A large sac of the former, marked *b*, produced by long-standing unrelieved retention of urine; a bougie lies in the small opening by which it communicated with the bladder.

distended gradually by the continued percolation into it of urine from the kidneys. Let your washing-out at least conform in some respect to that process. Never, under any circumstances, throw in more than two ounces; and even this quantity, for efficient washing, is better avoided. Proceed, then, as follows : You have a flexible catheter in the bladder ; have ready a four-ounce india-rubber bottle with a brass nozzle and stop-cock the nozzle long and tapering so as to fit a catheter of any size between Nos. 5 and 10, filled with warm water—say at 100° Fahr. Attach the nozzle gently to the catheter, and throw in slowly a fourth of the contents; let that run out, and it will be thick and dirty, no doubt; then inject another fourth which will be less so; again another, which will return clearer than the preceding; and the fourth portion will probably come away nearly clear. Now, these four separate washings of an ounce each will have been really more efficient than two washings of four ounces each; and you will, in obedience to my never-failing injunction, have reduced the amount of instrumental irritation to a minimum. Ten to one but the patient will regard your performance as soothing to his feelings. There are other methods of effecting the object, but this is the *principle* I want you to understand; and the mode of carrying it out which I have described is one of the simplest.

What if this washing-out has not accomplished all we wish? We may then, and often with great advantage, try medicated injections. Perhaps the best mild astringent, when the urine is alkaline and depositing phosphates, is the acetate of lead, in the proportion of one grain to four ounces of warm water, not stronger; to be used once a day. After this comes the dilute nitric acid; one or two minims to the ounce of water. Then you may try nitrate of silver in small quantity—certainly not more than one grain to four ounces to begin with, going up to about half a grain, or one grain at most, to the ounce. You may also use, especially where the urine is offensive, carbolic acid; one or two minims to four ounces is quite strong enough. Then there is a soothing injection well worth your remembering—viz. biborate of soda and glycerine. It may be used where there is no great occasion for an astringent, or it may be combined with one. The value of this for sore mouth suggested to me its use for an irritable bladder, and experience has confirmed my expectation. Here is my formula: Two ounces of glycerine will hold in solution one ounce of biborate of soda; to this add two ounces of water. Let this be the solution, of which you add two or three teaspoonfuls to four ounces of warm water. I arrange all these solutions for four ounces, because the four-ounce india-rubber injecting bottle already described is a convenient and portable instrument.

In circumstances of great pain, you may inject anodynes into the bladder if you please; but they are of little value. And you need not be afraid of the quantity; for the mucous membrane of the bladder appears to have no absorbing power, unlike the neighbouring tissue which lines the rectum.* And there, indeed, is your place for action, if spasm and pain greatly disturb the patient; a suppository of cocoanut butter, containing from half a grain to a grain of morphia, is often of the greatest service. Counter-irritants play a small part among our remedies; perhaps the best and safest is a hot linseed poultice, well sprinkled with strong flour of mustard, above the pubes. I cannot say much for croton

* Some one thought proper to question, in one of the journals, the accuracy of this statement relative to the effect of narcotic injections into the bladder, and even to caution my readers against relying on me too implicitly. It might have seemed otherwise almost unnecessary to say, that this statement was the result of very numerous experiments and observations ; and its object was of course to show that such injections were of small service, and, therefore, not to be recommended. My only reply to the critic was, to inject *four drachms* of Liq. opii sed. into the bladder of a patient with chronic cystitis, in one of my wards in Univ. Coll. Hospital, on four separate occasions, in presence of the students, who verified for themselves the absence of any sign of the presence of opium in the system. Subsequently, a dose (*by mouth*) of twenty minims, produced them all most notably.

oil, nitrate of silver, &c., there. Hot fomentations, in the form of bran-bags, hot flannels, &c., alleviate pain materially; so also hot hip-baths and the hot bidet.

Then there is a host of infusions and decoctions reputed to exercise a beneficial influence in cystitis. I will name some of them in what I think to be about the order of their value for the cases one commonly meets with : Buchu, Triticum repens, Alchimella arvensis, Pareira brava, and Uva ursi. Now, for the doses of these, your conventional tablespoonful is a miserably inefficient measure. Of the first, fourth, and fifth, give half a pint daily; of the second and third, a pint, that is of their infusions or decoctions, as the case may be.

The underground stem of the Triticum repens, or the common couch grass, was introduced some years ago by myself. Of this I will only say that it maintains its credit, and is undoubtedly very useful in many cases. For use, boil two ounces in one pint of water for a quarter of an hour ; the strained liquor to be taken by the patient in four doses in the twenty-four hours. It was a favourite remedy in the old herbals; and it formed the staple medicine against what was called "strangury," which, a few centuries ago, meant everything like pain or difficulty in making water, no matter what the cause; for the art of diagnosis then was in its earliest infancy. The "Parsley piert" (derived from " percer la pierre," and not a parsley or umbelliferous plant at all), or Alchimella arvensis, has proved in my experience an admirable remedy in obscure cases. Use it as an infusion : one ounce to the pint. Besides these there are the resins, which have a certain amount of influence upon the mucous membrane of the bladder; such, for instance, as copaiba, Venice turpentine, &c. You should not, however, give the dose which you would give in gonorrhœa. Five minims of copaiba, three or four times a day, in mucilage, often answers well. I may say the same of the oil of cubebs.

One word about alkalies. As a rule, no doubt, alkalies, in neutralizing highly acid urine, help to control chronic cystitis; an'd I like the liquor potassæ, as well as the bicarbonates, tartrates, and citrates, which appear to have more diuretic action, and to increase the quantity of urine, when you would rather avoid this action and lessen the frequency of micturition. The old combination of liquor potassæ and henbane, affirmed to be a union of incompatibles, nevertheless seems to me about the most valuable form in practice. I have no doubt that it is sufficiently correct that both henbane and belladonna are deprived of their specific activities when mixed with liquor potassæ. Chemically, I dare say that is so. But I am perfectly satisfied that this combination materially controls painful and frequent micturition in the complaint we are considering. Hence I have of late gone back to it, and for the reason stated.

Now as to acids. Remember that these are by no means the complement of alkalies in relation to the urine. Beware of the popular notion that it is possible to produce an acid reaction on urine by giving mineral acids by the mouth. By giving alkalies, you can make the urine neutral or alkaline to any extent you please, but you cannot do the converse with these acids. Yet I constantly hear it said, " The patient's urine is very alkaline; had we not better order acids ? " My reply is, " By all means; give an ounce or two daily, if you like, but it will not change the reaction of the urine." I have given these quantities, greatly diluted, of course, without the slightest effect on alkaline urine. No doubt mineral acids are useful in giving tone, and so do good; but don't prescribe them with the view of directly acting on the urine. The acids that do act on the urine are benzoic acid and citric acid, but you have to give so much of these that I do not know whether the remedy is not mostly worse than the disease. The benzoic acid, having some balsamic character, may be useful in some cases of chronic cystitis. The best way to give it is in pills, as it is not soluble in water. Three or four grains, with one drop of glycerine, is a good form; and you must give as many as ten or twelve pills a day if you wish to do any good. At all events, it is useless to give less than six; that would be twenty-four grains in the day. Lemon-juice has also an acid influence on the urine, and if it agrees with the stomach, may be taken in large quantity. But here is the important fact for you to remember: Surplus of acid in the urine is a constitutional error, and it enters the urinary passages at the kidney. It requires constitutional treatment, of the digestive rather than of the excretory organs, and mere alkaline treatment does but mask the acid, does not cure it. You have to remodel the patient's habits, control his diet, and take care that his liver and bowels act healthily and freely. On the other hand, persistent alkali in the urine is, in nineteen cases out of twenty, a *local* formation in the bladder. It requires local treatment, as by catheter and injecting-bottle, and not physic. Now and then you have alkaline urine, milky-looking, with amorphous phosphates, as a constitutional condition; but this is rare as compared with the cases I am now describing.

I shall close this lecture with some brief remarks on acute and chronic prostatitis.

Acute prostatitis occurs in different degrees of severity, and often comes first before the practitioner's notice, when it causes retention of urine by obstructing the neck of the bladder. How this emergency is to be met I have described at some length in the fifth lecture. The organ is often considerably swollen and very tender, and the inflammation may give rise to abscess in the substance of the gland, or adjacent to it; and the matter may burst either into the urethra, its most common course, or into the rectum.

Chronic inflammation of the urethra passing through the prostate, and more or less affecting the prostate itself, is a condition less generally known or recognized. Nevertheless, it is a common and important affection. We see it frequently, not always, as the result of obstinate gonorrhœa. I have already referred to it as the cause of symptoms resembling, more than any other malady, those of calculus in the bladder when mild in degree. Thus a patient of twenty or thirty years of age tells you that the following symptoms have rather gradually appeared:—Undue frequency of micturition; pain following the act, and felt in the end of the penis; occasionally a little blood seen with the last few drops of urine, which may be a little cloudy with muco-purulent deposit; a sense of heat and weight in the perineum and rectum; there is, perhaps, also some gleety discharge in the urethra. All these conditions are aggravated by exercise. You see he gives you a complete sketch of the symptoms of calculus; and how are you to distinguish them? By the history and by sounding. Thus, there is no history of the descent of calculus from the kidney, nor of gravel previously passed. But there is the fact of a chronic gonorrhœa resisting, perhaps, months of treatment. And if the patient shows no improvement, you must not decline to sound him. You do so, and find nothing, but that the prostatic urethra was very sensitive, and you make him worse, perhaps, for a day or two.

What is to be done? First and foremost, as a rule, abjure all instruments, which, in most cases, can only do mischief. Treat it as you would a chronic inflammation of the ear or eye—*i.e.* blister an adjacent surface ; make a small blister every four or five days on either side of the raphé of the perineum, by applying with a brush the Lin. epispast. of the Pharmacopœia, not so freely as to distress him or prevent locomotion, and keep it up for four or six weeks. I have found the best results from this method, combined with a tonic medicine and regimen, and you will find the patient himself gladly exchanging the dull weary aching in the perineum for the smart of the blister, and cheerfully noticing how the former gradually subsides under the influence of the latter. In exceptional cases, where chronic gleet is a prominent symptom, the application of a solution of nitrate of silver, not more than five or ten grains to the ounce of water, to the prostatic urethra, may be very serviceable.

I shall, in my next lecture, proceed with diseases of the bladder.

LECTURE XI.

DISEASES OF THE BLADDER: PARALYSIS; ATONY; JUVENILE INCONTINENCE; TUMOURS.

GENTLEMEN,— On two occasions recently a patient has been sent to my ward affirmed to have "paralysis of the bladder;" such, at least, is the statement that accompanied them here. On examining one of them we found a not unhealthy-looking elderly labouring man, from whom, by much questioning, we elicited the following facts: That he was nearly sixty years of age; that he has passed his water much too frequently for four or five years; that he was much disturbed at night to do it, although lately it has come away without his knowledge during sleep; that when he makes an effort at work the same thing often happens; that the stream is weak, falling almost perpendicularly; that he has "no particular pains," but is not so strong as formerly, having become much weaker of late; and that for the last few months the urine has been cloudy and has had a disagreeable smell. With all this his ordinary functions had been fairly performed, and he had followed his daily labour until three weeks ago.

The man was desired to unfasten his dress; as he did so you remarked a urinous odour, and that certain cloths, which did duty for an india-rubber receptacle,—a luxury beyond the means of our patient,—were wet with the secretion. Two conditions only could cause this unhappy state of things: either the bladder was incapable of doing its duty as a reservoir, and permitted the urine to escape as fast as it entered from the ureters, or it was unable to expel its contents, so that it was over-distended by them, the surplus oozing out, or being forced out, in the manner described.

Now a glance of the eye might have nearly sufficed to settle this question. I pointed out a marked protuberance above the pubic symphysis; and after placing the patient on his back, the dulness by percussion corresponding with that spot, and the clear bowel-note all round, diminished the doubt that this was a collection of fluid, if any such doubt yet existed. Still this was not quite all that it was necessary to know; it was just possible that the swelling might be a solid tumour of the bladder, occupying its proper space and much more, and so destroying its function as a reservoir. To the hand, however, the protuberance was clearly made up of fluid ; but even this is short of absolute demonstration, for the most practised hand has been known sometimes to "lose its cunning," or to have found a too deceptive quality in the object handled. Finally, you saw that a well-curved gum catheter glided into the bladder, and upwards of 40 ozs. of somewhat stale urine flowed off. I then examined the prostate, and found no very obvious enlargement.

Now, was this a case of "paralysis of the bladder?" Certainly not. His history showed that he had had no seizure of any kind, and I beg you to understand that without some change in a nervous centre there is no paralysis of the bladder. Recall, if you please, what I said in my fourth lecture on this subject. This term is applied, or rather mis-applied, every day to such cases as the one before us, and with the result not merely of masking the true pathological state, which ought always, if possible, to be indicated by a nosological term, but of misleading the inquirer, since it indicates a condition which by no means exists.

What, then, is the defect or disease occasioning the symptoms in this case ? Probably, atony of the bladder. I will speak presently with more precision. The bladder fails to expel its contents in the two following conditions: either a growth from the prostate, by no means necessarily large, obstructs the neck, so that the natural power of a healthy bladder—which may, moreover, have been reinforced for the difficulty, by hypertrophy of its fibres rendering it stronger —cannot propel the urine by or over the obstruction into the urethra ; or these muscular fibres of the bladder are so enfeebled, or even atrophied, that their propelling power is lost or greatly diminished, and the organ becomes a thin flaccid bag, and can exert no expulsive force upon its contents.

The two conditions sometimes coincide. We may have hypertrophy of the bladder following obstruction from stricture of the urethra. On the other hand, with obstruction from the prostate, the vesical coats sometimes extend and become thin. But atony may be produced without any disease of the prostate, and then mainly from the subject of it having been placed in circumstances which obliged him to retain his urine for a too considerable period, so that the bladder became over-distended, and has failed subsequently to regain its tone. Unfortunately, a single error of this kind will sometimes produce an irremediable atony.

Now, on further questioning our patient, we did not find that he could recall any such instance, nor that he ever formed the habit of permitting such over-distension. Neither did the affection occur suddenly ; on the contrary, the symptoms appeared gradually. What is still more significant is, that they occurred just at that time of life when prostatic hypertrophy commences, if it appears at all.

Still, the prostate was not obviously large on examination by the bowel. We arrive, then, at the following conclusions :—That this man has some enlargement of the prostate, which, though not obvious in the rectum, consists in a small nipple-like projection of the median portion, occluding the neck of the bladder, and that, from the size of the bladder, as just now demonstrated by percussion and by its contents, its walls are thin, and have lost their contractile power ; in other words, are in a condition of atony.

I think there is no escape from these conclusions, and I beg that you will not only never permit yourselves to allude to this condition as " paralysis," but that you will protest against so loose and improper a use of the term when you hear it thus misapplied by others. Now, true paralysis of the bladder occurs from injury to the spine, and also as one of that large group of symptoms which result from disease in the cerebral or cerebro-spinal centres. You find it associated with slightly unsteady gait, with impaired articulation, or with some of the slighter signs of such central mischief, as well as with those which are most obvious ; and I have even found it persisting after all other signs have nearly—I cannot say quite—disappeared.

In all cases, as in that of this patient in the ward, it is essential to empty the bladder by means of the gum catheter three or four times daily, to remove the urine completely, and at least to offer the possibility of re-acquiring power to the muscular coat, which does not exist so long as that coat remains constantly distended by the retained urine.

Next, for pure atony and for slight paralysis, uncomplicated with prostatic enlargement, a little aid may be sometimes afforded to the patient through the agency of electricity, by cold douches and injections, and by tonics ; but less advantage is to be derived from these means, in my opinion, than some have appeared to believe, although I by no means say they are not sometimes serviceable. I have seen an increase of expulsive power attained rapidly during the daily application of an electro-magnetic current to the bladder, and in the following manner:—To one pole the ordinary handle and moist sponge are attached, which is placed over the lumbar vertebræ; an elastic bougie, containing a conducting wire, and tipped with metal, is attached to the other pole, and is introduced into the bladder. A weak current is set going, and its effects watched. Thus a slight sensation only is to be produced. Move the bougie about gently in contact with the walls of the bladder, the urine having been just withdrawn; and, finally, let it rest a little in the neck of the bladder, where greater discomfort is felt: in all, allowing the current to pass for eight or ten minutes before withdrawing.

A very different condition to that just described sometimes results after severe lesion, and also after local injury—viz. inability on the part of the bladder to act in any way as a reservoir. In this unfortunate situation the urine leaves the organ by the urethra as fast as it enters by the ureters. This is complete incontinence, in the true sense of the term. Little else than mechanical contrivances are of any avail. And these consist in making a reservoir outside instead of inside the body; one which can be emptied at the patient's will. Happily, such cases are very rare.

But there is a partial incontinence which is very common, and which is, moreover, amenable to treatment. You will be consulted by an anxious mother who, bringing her boy or a girl, of any age below puberty, and occasionally above that period, tells you that every night, or nearly so, this young person wets the bed. Examples of this are frequently seen

M

in my out-patient room. You know that, in a child with a
busy excitable brain, muscular movements occur during
sleep of a much more active character than those which
usually occur in the adult, or in children of a more placid
temperament. Anything up to somnambulism may take
place during sleep in a child whose *physique* is weak, and the
slave of a restless, ceaseless activity of mind; and micturition during sleep often occurs in connection with this state.
Clearly, however, not only in such cases ; for in some very
dull and stupid children, in whom intelligence appears to
be below the average, the same thing may happen. And it
must also be admitted that there are cases which do not fall
into either of these classes. For such unfortunate patients
all sorts of remedies and all kinds of management are adopted,
including even a periodical employment of the birch—a
species of "cytisus" which I trust you will never admit into
your own therapeutic scheme. Depend upon it that " pun-
ishment" for this form of youthful frailty will not answer ;
and whatever of strength to the moral faculty may be communicated in obedience to the ancient injunction not to spare
the rod—a question beyond our province to discuss—do not
regard it as binding on us who practise the healing art.
The child's attendants often lose patience at the perpetual
recurrence of the disagreeable infirmity, and believe it to be
the result of wilfulness or of carelessness. I have seen much
cruelty practised, even by the nearest relatives of these unhappy offenders. Give it no countenance whatever.

Now for treatment, briefly, and as much as I can on general principles. For class the first, you will cultivate the
physical side of life ; remove as much as possible the sources
of over-mental stimulation ; strengthen the constitution
through the agency of diet, country air, or sea-bathing, if

possible, and give steel wine and cod-liver oil. For class the second,—those of torpid and deficient intelligence,—you must show the importance of developing mind. Endeavour to call the will into play as much as possible, and enlist it to aid you in preventing the act. These are the children who are usually ill-treated; instead of which they must be made sensible of the degradation of the habit, so as to get a stimulus for volition in relation to it. Here remedies which act specifically on the organs are most appropriate. First and chiefest is belladonna; which apparently paralyzes the expulsive muscles of the bladder and the sensitiveness of the organ at the same time. Thus in elderly people, who have feeble power to expel the urine, a dose often produces complete retention, and that for some time, unaccompanied by consciousness of inconvenience from it. You give small doses, say of the tincture, in the afternoon and evening, increasing gradually when it may be necessary, and if the bladder is thus made to retain the urine all night for a few weeks, on relaxing the dose gradually, the habit of retaining is found to be formed and to persist. This remedy is so excellent a one, that it has almost superseded blisters to the sacrum and such counter-irritants. After it, nux vomica may be tried. Then, for confirmed cases, after the failure of other treatment, especially for those who have arrived at puberty, or thereabouts, a mild caustic solution to the prostatic urethra—say ten grains to the ounce of water—has answered in my hands; to be repeated with a stronger solution if necessary.

In all cases inquire carefully for derangements of digestion, in all its stages, from primary stomach symptoms to worms in the lower bowel—not unfruitful in their adverse influence. Of course you will take care that the child has not too much fluid, nor takes too much of solid food containing much water

M 2

in its composition, during the latter third of the day; and that it is taken up to pass water late in the evening, when its attendants go to bed.

A short sketch of Tumours proper to the bladder will finish this part of our subject. Of course you are to understand that all those outgrowths from the prostate which come under the head of hypertrophy, since they are more or less composed of structure identical with, or very similar to, that composing this organ, however much they may project into the cavity of the bladder, are not to be included in the present class.

Tumours proper to the bladder are of rare occurrence. I wish you to know what they are—what it is possible you may meet with, so that you may have the chance of recognizing an example if it falls in your way. As with those in other parts of the body, they are classified according to the amount of force which they manifest to invade surrounding structures, or to reproduce themselves elsewhere. Thus, first, we have simple fibrous growths, chiefly in the form of polypi, springing from the walls of the bladder, and wholly unassociated with the prostate; the rarest of all forms—in short, exceedingly rare, known to me personally only in museums.

Secondly, there is the "villous or vascular tumour" of the bladder, miscalled "villous cancer;" for it has no invading or reproductive power beyond the organ in which it arises.

Thirdly, there is epithelioma—the lowest type of malignant formation, and slowest of development.

Fourthly, true scirrhus occurs in the walls of the bladder; and, much less commonly, encephaloid.

Now, putting the first form out of the question, it may

be said in general terms of all the rest, that the single and most certain characteristic of the presence of a tumour is *persistent vesical hæmaturia*, no calculus or other obvious cause existing.

But never arrive hastily at the conclusion in your own mind, nor even too readily admit the suspicion that tumour is present : for you will remember, in the first place, that it is exceedingly rare, compared with other causes of similar symptoms ; and, secondly, that there are no positive signs, or almost none which can be so regarded, of its existence. It is only by a long process, and after much careful watching of any case, and consequently when the disease is in a somewhat advanced stage of development, that you can, *per viam exclusionis*, conclude with some reason that a tumour is present. The symptoms are almost identical with those of calculus, and the patient is certain to be sounded more than once or twice before tumour is even suspected. Suppose, then, that you have verified the absence of stricture, of prostatic enlargement, of calculus, of primary renal disease, and you are at a loss to know why your patient passes water with great frequency and pain, the secretion containing more or less muco-pus, and having blood in it often or continuously, the quantity of which is increased by exercise,—you direct your inquiries to the existence of vesical tumour. And you will proceed thus : First, you will introduce into the bladder the short-beaked sound, and with a finger in the rectum you will carefully explore the thickness of the structures intervening between the finger and the sound. Next, the sound being still in the bladder, you will not find much difficulty, provided the patient is thin, in gaining some information of a like kind by palpation above and behind the pubic symphysis. Further, by

movements of the sound itself, you may detect a hard mass
of scirrhus on either side of the vesical walls, the sound not
turning over readily to the left or right, as the case may be.
You will not discover a villous tumour thus, for it is much
too soft, and will elude the most delicate traversing of the
cavity which can be achieved.   Even an epithelial growth,
which is usually wide in its base, of flocculent surface, and
sprouting into the cavity, although not very luxuriantly, is
so deficient in induration as not to be readily discoverable.
It scarcely destroys the flexibility of the vesical coats, which
is the fact you have to ascertain.

Next, you may search for enlarged glands in the iliac
regions, but they are palpable only in advanced cases of
scirrhus, and you will obtain such light on the subject as a
search for cancerous growths in other parts of the body
may afford you.   Thus, not long ago, I had my diagnosis of
cancer of the bladder in the case of an elderly patient made
certain by the appearance of a secondary growth, springing
from the cranium.

Again, you will repeatedly and carefully examine the de-
posits in the urine for the appearance of organic materials
cast off from the growth, which may serve to indicate its
nature.   Thus I have detected under the microscope the
peculiar structure which the processes of the villous tumour
present to the eye.   But what of the cells of epithelioma,
and what of "cancer-cells?"   I am compelled to resign to
others—and I am well aware that several writers on this
subject have proclaimed the value of microscopic examination of the urine in vesical cancer—the good fortune of
identifying malignant disease by this means.   First, suppose
you have caught your "cancer-cell," are you prepared to
swear to its identity?   As students, I will assume you have

examined, say a few hundred specimens of urine, not many, but enough, at all events, for my purpose, and that you have, therefore, not a little perplexed yourselves, if you are of an inquiring turn, with all that fruitful progeny of cells, epithelium of different parts and in all stages of growth, &c., which are desquamated in health, but especially under the influence of any morbid action in those passages, and which appear therefore in the urine. Some of the best "cancer-cells" I ever saw in my life were collected from a patient's urine, and placed under the microscope by an eminent microscopic observer, for the purposes of a very important consultation, at which I assisted. After a careful examination of the patient, I admitted the beauty and perfection of the microscopic observation, but on larger grounds denied the existence of cancer in the bladder; and the ultimate result, happily for the patient, confirmed that view, and disproved the cell. Most valuable as is the microscope in this great class of maladies, ranking next, and very near to, the sound itself, never let it obscure for you those broad features of the case which are to be determined by the unassisted eye and touch, as applied to the body, and to the urinary secretion itself through the means of reagents. But when you find, as sometimes happens, distinct masses of soft, almost semi-translucent structure of considerable size, passed by the urethra at micturition, and discover on examination that these evidently consist of rapidly formed cell-growth, the cells of large dimensions, and containing two or three nuclei, the observation will go far to confirm your suspicions of cancer aroused by pre-existing symptoms.

Lastly, you will, in endeavouring to determine the particular kind of growth, observe the nature of the hæmorrhage and the character of the pain. In malignant disease, the hæmorrhage is irregular in its occurrence, long intervals being sometimes observed in which no bleeding appears; and when it does take place it is often in considerable quantity, and it may persist for some time; moreover, the blood is usually of a florid colour. In villous growths the urine is generally more persistently coloured, and presents a reddish tint, resembling the juice of raw meat—not dark or smoky in hue; occasionally, however, rapid and considerable hæmorrhage takes place. The pain of cancer is more constant and severe than that of villous growth, the latter not being necessarily accompanied with great pain, unless obstruction to the outflow of urine is produced by the tumour.

What shall I say here respecting the treatment of vesical tumour? That it must be shaped according to the existence or predominance of certain symptoms, which may be regarded as three in number: hæmorrhage; painful and frequent micturition; retention of urine.

For the treatment of hæmorrhage, I have little to say for the internal astringents; I mean those administered by the mouth, as gallic and tannic acids, lead, alum, matico, &c., the last-named being perhaps the best. I know nothing so valuable as injections into the bladder of nitrate of silver, commencing with a very weak solution. But for the local treatment of vesical hæmorrhage, never forget how essential are delicacy and gentleness in all your manipulations, otherwise you will provoke hæmorrhage rather than restrain it.

For alleviation of pain or frequency of micturition do not spare opiates—trying any form, or all forms in turn, until you find that which most assuages it and least disturbs the digestive organs. Give them by mouth, subcutaneously, or by suppository. Never mind how much, in order to act effectually. It is not a question of saving life, but a question of mitigating that most frightful of human miseries—prolonged, continuous, severe bodily suffering; and this for a patient whose doom is certain, and to whom life has come to be for the most part a dire calamity. While you are bound, therefore, on the one hand, jealously to guard life, I hold that you are equally responsible, on the other, that it shall be rendered fairly endurable. I confess that I have felt sometimes almost indignant at the sight of a poor fellow-creature, worn out with anguish, praying for death, who, thanks to a well-meaning but weak timidity, is permitted only such small comfort as fifteen or twenty minims of liquor opii, or of a solution of morphia, once or twice in the twenty-four hours can afford.

For the relief of retention, such catheterism must be applied as the case requires, whether periodically or continuously, as the comfort and the exigencies of the patient render desirable.

I shall close this course with a lecture on "Hæmaturia," which will permit me to glance at most of those subjects which have not yet come under our consideration.

LECTURE XII.

HÆMATURIA AND RENAL CALCULUS.

GENTLEMEN,—I propose to-day to complete the programme I originally designed for this course by considering a phenomenon of common occurrence known as Hæmaturia.

Let us define the term. What is Hæmaturia? The outflow of urine containing blood in admixture. Thus bleeding from the penis at other times than at that of micturition is, of course, not hæmaturia. Bleeding coincident with micturition from chordee, or operation, or from any known injury in the course of the urethra, is also not to be included. The blood then usually issues by the side of the stream of urine, and is only partially mixed at its line of contact, or it may follow rather than accompany the urine.

Hæmaturia, then, is a symptom. Its presence is, in all cases of urinary disease, to be sought. Hence the inquiry forms one question—the third—of the necessary four which I instructed you always to ask in forming a diagnosis. Here is a glass of urine, evidently containing an admixture of blood. What is the source of it? Now, it is often not an easy thing to state at once what point of that long and complicated organic apparatus, which commences in the Malpighian corpuscles and ends at the external meatus, is the source of the blood in question. Sometimes it is exceedingly difficult to define its source. Thus it is that in medicine you will often find some symptom, the pathological cause of which is not very obvious, getting a specific name, which comes at length to denote a distinct disease; and just as you will be asked, as I told you the other day, what is good for dropsy, you may also be asked what is good for hæmaturia.

Now, the consideration of this question, besides affording us new material for inquiry, will bring us over some of the ground we have already travelled together. I don't regret that —for your sakes, I mean. It will stand in the same relation to the past as the arithmetical "proof" does to the already worked sum. 'It is in some respects a synthetical operation following an analytical one. When, therefore, you see a specimen of urine containing blood, you will, as a matter of course, make a rough mental note of the proportion of blood present, and you will mark the colour. And as you can count on your fingers the ordinary sources of blood, these will pass rapidly in review at the same time. Let us name them as follows :—

1. The kidneys, where it may be from diseased action, more or less temporary, as inflammation; or from disease more or less persisting, as degeneration of structure; or from mechanical injury, as from calculus there, or by a strain, or a blow on the back. If the hæmaturia is the result of inflammation, there will be general fever denoting its presence; if produced by slow organic change, there will be the history of failing health, and probably urine changed in quality otherwise than by the mere admixture of blood. Where blood is in very small quantity, as it will naturally be at times, note the character of the urine proper—whether of low specific gravity, pale, with albumen in greater proportion than blood or pus will account for; perhaps renal casts may be found ; and look out for dropsies in any degree. In both the preceding forms, if blood is present, it will give the smoke tint to the secretion. Perhaps it may be affirmed that such

urine, associated with very little if any local pain, is more likely to come from the kidney than elsewhere. In malignant renal tumour, blood may be large in quantity at times; the rapidity of growth and the size attained, are the marked characteristics of the disease. If mechanical injury be the origin of hœmaturia, there will be the history of a blow or strain; or there may be the signs and symptoms of renal calculus, of which more detail presently.

2. Then, putting aside the ureters, you will remember the bladder as the second source of hæmorrhage; and here it may be due to severe cystitis, calculus, or tumour. The first is obvious enough from muco-pus in the urine, and through other signs; while the second may well be suspected by the symptoms, and its presence realized by the sound. Here the hæmorrhage is usually florid, and in proportion to the patient's movements. But the third condition—namely, that the hæmorrhage arises from tumour—is not always so readily to be affirmed. As a rule, however, blood from such a source is larger in quantity than from stone, and may be associated with less of muco-pus. If the tumour is malignant, it may be felt by examination, and the pain is often severe; if villous, it gives an even pale-red tint for days together to the urine; and in both cases the blood is florid, unless it is long retained in the bladder, when dark sanies, like coffee-grounds, results.

3. In hæmorrhage from the prostate, the third principal locality or source, the same thing occurs, if the organ is hypertrophied and the blood is retained; but here the age of the patient, and the ascertained condition of the organ from the bowel, aid the diagnosis. A slight appearance of blood mixed with the last few drops of urine is not a rare occurrence in chronic prostatitis.

4. When bleeding arises from stricture of the urethra, the patient's history and the cause of the bleeding, almost always instrumental, leave no room for doubt. From the use of instruments also in the bladder, hæmorrhage sometimes arises. Then it is not to be forgotten that occasionally blood is found in the urine as the result of violent diuretics, from purpura, in fevers, and in a hæmorrhagic diathesis.

Now for the treatment of hæmorrhage. When you have determined that its source is above the bladder—that is, in the kidney or in its pelvis, probably the first and most influential remedial agent is rest in the recumbent position. Whether from a lesion affecting the intimate structure, or from the mechanical irritation of a calculus in any part of the organ, rest is the first and the essential condition. The patient is, moreover, to be maintained in as cool and tranquil a state as possible.

It is in renal more than in any other form of hæmaturia, perhaps, that direct or internal astringents or styptics are useful. I shall do no more than name those which are most commonly used—namely, gallic and tannic acids; lead and turpentine; equal to them is, I think, the infusion of matico, say in doses of two ounces every two or three hours. The tincture of iron, and also sulphuric acid, may sometimes be taken with advantage.

It is, however, in cases of severe hæmorrhage from the bladder, or more commonly from an enlarged prostate, that active and judicious treatment is necessary. You will be called sometimes to a patient whose bladder is distended with coagulated blood, or who is passing frequently a quantity of fluid in which blood is the predominating element. Usually this has arisen from some injury inflicted by the instrument, although it may also be from tumour of the vesical walls. Here you will keep the patient on his back, and forbid the upright position, or any straining, so far as you can prevent it, in passing water. To this end give opium liberally, to subdue the painful and continued action of the bladder. Apply cold by means of bags of ice to the perineum and above the pubes. Better still, introduce small pieces of ice into the rectum. Do not use an instrument if it is possible to do without it. There is a great dread in some people's minds about the existence of a large coagulum in the bladder. I have even known a bladder to be opened above the pubes by the surgeon, for the mere purpose of evacuating a mass of clotted blood. Leave it alone: it will gradually be dissolved and got rid of by the continued action of the urine; while if you are in haste to interfere, and are very successful in removing it, you will succeed also most probably in setting up fresh hæmorrhage. The bleeding vessels have a far better chance of closing effectually if they are not subjected to mechanical interference. Meanwhile, support the patient's powers by good broths, &c.

But it sometimes happens that hæmorrhage occurs in a patient who has long lost all power of passing urine except by catheter. This is a very different position. Here the coagulated mass which fills the bladder must sometimes be removed, or no urine can be brought away. Thus, you introduce a catheter, and none appears, for the end of the instrument passes into a mass of coagulum, and nothing can issue. Sometimes sufficient may be removed by attaching to a large silver catheter a six-ounce syringe or a stomach-pump. Clover's lithotrity apparatus has answered remarkably well with me in two or three instances. Let me caution you, as a rule, never to inject styptics into the bladder in these circumstances; the irritation does more harm than good.

The only exception I have ever found serviceable has been in injecting a little iced water, or iced solution of matico, immediately after removing the clots from the bladder.

In passing to another subject, I beg to call your attention to a glass before you containing some urine of a dark and somewhat unnatural tint. Let us interrogate the case of the patient from whom it came. In obtaining this specimen I took care that he should first pass about an ounce into a separate vessel, to clear his urethra—a precaution always absolutely necessary to avoid error, as I have before warned you—and the remainder into this. It is less translucent than average healthy urine is, and has a deeper colour. The hue is not read, but slightly orange, with a dirty brownish tint, commonly and very well distinguished by the word "smoky." That tint denotes blood to an ordinarily practised eye. Why is it not red ? Because blood, after a certain term of contact with urine, loses its red colour and becomes brown; and you see it in that condition, or according to the quantity, producing any depth of hue from this up to that of London porter. Put it under the microscope, and you will find plenty of blood-corpuscles. We get this broad principle, then, to start with : bleeding from the more distant parts of the urinary system, unless in very large quantity, will almost certainly make the urine brown, while urine which contains red blood has almost certainly issued recently from some source in the bladder, probably at or near its neck, these being the more common sites of vesical hæmaturia.

In the case before us, then, we proceed easily and rapidly to eliminate many of the sources of bleeding by physical exploration, and by the account which the patient gives of his sensations. He makes a good stream when a fair amount of urine has accumulated in his bladder, but this does not often happen, for he passes it every two hours, or less, in the day, not so frequently at night; no straining is necessary. Pain in the course of the urethra is experienced during and after micturition ; not severe. He is uneasy about the pelvis and loins on taking exercise, and more blood passes afterwards. He is somewhat emaciated, and so presents a good condition for examination by the hand. He is subject to variation in the intensity of the symptoms, having now and then attacks of a few days' duration, in which they are aggravated, and he dates their commencement in an attack which occurred seven years ago. His age is forty-five years. He has never passed gravel. He is much less robust than formerly ; his digestion is not good. A full-sized bougie passes easily into the bladder; no stricture; hypertrophy of prostate at his age not possible. On sounding, he is manifestly more tender than usual; nothing is felt, nor any deviation from the natural condition by simultaneous examination by the rectum. Palpation of lower part of abdomen shows nothing. Arriving close under the last ribs of left side with one hand, the other pressing firmly on left renal region, he flinches unmistakably ; that is the spot, he says, where he feels pain at times and on movement ; right side, nothing. We examine his urine : sp. gr. 1018, acid, small brownish deposit on standing ; under microscope, blood-corpuscles, some pus-corpuscles, epithelium, no crystals, no casts; albumen, a little; corresponding with organic matters present.

What is the seat of the lesion in his case? You say, perhaps the bladder : we found it tender to the sound, and it acts with undue frequency. Yet remember this is by no means evidence of any primary morbid change there, such conditions constantly accompanying diseases affecting primarily the kidneys or the upper part of the ureter. Much more probably the kidney. The manifest local tenderness, the repeated attacks, the impaired health, the history, the absence of all the more common causes of cystitis in any form, point to the left kidney as the seat of mischief. The absence of albumen and renal casts—a fact of not much weight, although their presence is of the utmost importance—lead us to believe him free from organic changes affecting the renal organs. I conclude that his left kidney is the seat of calculus, although he has never passed one, and has at present no crystalline deposit in his urine—a fact by no means essential to the diagnosis; and that this calculus is the source of the blood and pus found in his urine.

It is sometimes not easy to say what kind of calculus exists in these cases, of which this is a fair type. When any calculous matters have been passed which can be examined, or when the crystalline deposit in the urine is constant, the inference is pretty clear. Add to this that the probability in any case is strong in favour of uric acid, from its known frequency of occurrence—taking large numbers, say at least fifteen to one as compared with oxalate of lime.

For treatment: Alkaline diuretics and diuretic vegetable infusions, before named, for a period of time; attention to the digestive functions and to that of the skin, for the kidneys are probably working too much vicariously for some other function acting lazily; moderation in highly nitrogenized food; mild alcoholic drinks, perhaps in most cases permitting only a light and mild Bordeaux. Of all medicinal remedies, perhaps none are so valuable as mineral waters, especially those which have sulphate of soda, largely diluted, as the main ingredient. Take Carlsbad, Friedrichshalle,

N

and Marienbad as examples. For two well-known remedial
agents, which are very popular, each among its class, I am
bound to tell you, I have very small esteem. Here, in town,
it seems to me that every man advises his neighbour, and on
every pretext, to drink Vichy water—advice which is cheap,
and of which the value in most instances by no means
exceeds the cost. In the country, where perhaps the fairer
sex more usually dispense similar aid to their suffering
neighbours, the prescription is mostly gin-and-water. Of
the first, or natural product, which is a strong solution of
carbonate of soda, I must say that, if not absolutely injurious, it is at least greatly inferior to potash ; and of the
second, or artificial one, that it is about as serviceable to
the kidneys as a pair of spurs to a jaded horse—makes him
travel for a time, but takes it out of him in the long run.
For the paroxysms of severe pain which denote the passage
of a renal calculus, you will find hot hip-baths, prolonged
or often repeated, of the greatest service ; the temperature
may be increased to anything the patient can bear. Large
doses of opiate, also, to assuage severe pain ; and abundance
of barley-water, potash-water, or the like, for drink.

I shall here, by way of episode, refer to a mode of determining the true characters of a patient's urine, which is of
extreme value in some doubtful cases—a mode which has
never to my knowledge been recommended or practised, and
which I have systematized for myself. I have already told
you how essential it is to avoid admixture of urethral products with urine, if you desire to have a pure specimen. It
is sometimes quite as essential to avoid its admixture with
products of the bladder. And I defy you to achieve an
absolute diagnosis—by which I mean a demonstration, and
never be satisfied with less, if it be practicable—in some few

cases, without following the method in question. When, therefore, it is essential to my purpose to obtain an absolutely pure specimen of the renal secretion, I pass a soft gum catheter, of medium size, into the bladder, the patient standing, draw off all the urine, carefully wash out the viscus by repeated small injections of warm water (before shown to be rather soothing than irritating in their influence), and then permit the urine to pass, as it will do, *guttatim*, into a test-tube, or other small glass vessel, for purposes of examination. The bladder ceases for a time to be a reservoir; it does not expand, but is contracted round the catheter, and the urine percolates from the ureters direct. You have, indeed, virtually just lengthened the ureters as far as to your glass. And now you have a specimen which, for appreciating albumen, for determining reaction, and for freedom from vesical pus, and even blood, and from cell-growths of vesical origin, is of the greatest value, and has often furnished me with the only data previously wanting to accomplish an exact diagnosis. Mind never to be satisfied to guess at anything; make, very cautiously, if you will, your provisional theories about a doubtful case—indeed, the intellectual faculty will do this constantly, and without reference to the will—but arrive at no conclusion, take no action, except so far as you are warranted by facts.

I have reserved these few words to the last, as the most important. The first words I uttered in this course were designed to convey to you my strong sense of the importance of acquiring the habit of making an accurate diagnosis, and a rapid one, if possible. My last shall be to express once more the same conviction. Not because I undervalue the subject of treatment, but precisely with the opposite view; being anxious, above all things, that you and I should afford some essential service to those who have confided to us the care of their maladies. I adjure you to spare no pains to obtain the most complete knowledge of the complaint itself, since it is the only mode of arriving at a knowledge of what will be sound and efficient treatment.

I beg to thank you for the extreme attention and assiduity with which you have followed me during this course, and to assure you that such a manifestation on your part has rendered our meetings for these colloquial discourses some of the most agreeable relaxations which have fallen to my lot, to vary the routine of an anxious and very active professional life.

THE END.

Woodfall and Kinder, Printers, Milford Lane, Strand, London, W.C.

www.ingramcontent.com/pod-product-compliance
Lightning Source LLC
Chambersburg PA
CBHW021711210326
41599CB00013B/1611